Collaborative Process Improvement

Collaborative Process Improvement

With Examples from
the Software World

Celeste Labrunda Yeakley
Jeffrey D. Fiebrich

WILEY-INTERSCIENCE
A JOHN WILEY & SONS, INC., PUBLICATION

The authors are grateful for permission to include the following previously copyrighted material:

Page 131, reprinted with permission from *Quality Audit Handbook,* 2nd ed., ASQ Quality Press, © 2000 American Society for Quality.

Page 52, reprinted with permission from *Critical Shift: The Future of Quality in Organizational Performance,* ASQ Quality Press, © 1999 American Society for Quality.

Page 107, reprinted with permission from "How to Speak the Language of Senior Management," *Quality Progress,* © 2003 American Society for Quality.

Page 19, reprinted with permission from Hickman, C., Connors, R., and Smith, T., *The OZ Principle,* © 1994 Prentice-Hall.

Page 21, reprinted with permission from Blanchard, K., and Hersey, P., *Management of Organizational Behavior,* © 2000 Printence Hall.

Page 91, reprinted with permission of the publisher. From *The Blind Men and the Elephant, Mastering Project Work—How to Transform Fuzzy Responsibilities into Meaningful Results,* © 2003, D. A. Schmaltz, Berrett-Koehler Publishers, Inc., San Francisco, CA. All rights reserved. www.bkconnection.com.

Published by John Wiley & Sons, Inc., Hoboken, New Jersey.
Published simultaneously in Canada.

For general information on our other products and services or for technical support, please contact our Customer Care Department within the United States at (800) 762-2974, outside the United States at (317) 572-3993 or fax (317) 572-4002.

Wiley also publishes its books in a variety of electronic formats. Some content that appears in print may not be available in electronic format. For information about Wiley products, visit our web site at www.wiley.com.

Library of Congress Cataloging-in-Publication Data is available.

ISBN: 978-0-470-08460-1

10 9 8 7 6 5 4 3 2 1

I urge all,
"Quay til"!

Contents

List of Figures

Preface

BEFORE YOU STEP INTO YOUR WORLD

Today's economic climate makes it more important to work smarter. Dedicated Quality personnel are stretched to the limit, and time constraints force greater emphasis on effective process improvements. Yet building the "Quality Department" seems to be the lowest priority in many companies. With overwhelming amounts of work to do and no resources—at least any incremental resources—what is the Quality worker to do? The battle cry to "do more with less" has become "work hard, work smart."

This book offers a framework that we deployed at a medium-sized firm that specialized in software applications. With over 40 years of experience in the software quality industry to draw from, we built a lean strategy for improving processes than can be applied in any number of industries. We formed a strategy around developing software quality engineering advocates on each of the projects using existing personnel. Moreover, the approach provided in this book can net any organization gains to the bottom line via dynamic process improvement activities.

This book also addresses the use of process champions to perform activities at the organizational level. We found that by using advocates and champions, a Quality Department could extend its reach. In fact, engaging non-Quality personnel in process watching can be rewarding not only to the quality of the end product, but also to the overall development process, and will result in an increase in organizational effectiveness.

To ensure effective execution, a Quality Manager must be prepared to:

- Establish advocates and champions
- Create a training program
- Embrace continuous improvement
- Communicate goals effectively
- Manage by influence
- Analyze and promote process improvements using metrics, evaluations, and rewards

One question you may ask about this collaborative process improvement approach is, "How is collaborative process improvement different from Six Sigma efforts?" The collaborative process allows us to transcend boundaries of functional processes, which are inherent not only internally, but also between the company worker and the consumer. Perhaps an exploration of what makes a process fully defined comes with the realization of what reigns and throttles the flow of the process. For instance, Six Sigma, business process improvement, and like initiatives tend to focus around a single team with a limited number of members and a specific contract to improve on a targeted goal. Although these efforts are important and can certainly result in process improvements, they can tend to leave out the critical masses while focusing on the "critical few." In other words, process-improvement efforts can fail because the people doing the work are not invested in the prescribed solution to their problems. These problems fester and grow into bigger problems, and prolong the pain of a stagnant process. Working teams tend to remain detached from the improvement implementations and you hear comments like, "Here we go with another flavor of the week; maybe if we just ignore it, it will go away like all the other big ideas they've tried to push on us." We are not saying that big programs are doomed to failure; we are just saying that there are certain processes that you can ingrain into the web of your organization's culture via collaborative process innovation that will provide a fertile ground for innovative thinking.

For these sorts of improvements (e.g., standards guidelines, process frameworks), a Quality Manager must engage the entire organization in process improvements. Most importantly, some companies can perform collaborative process improvement with no incremental increases in resources. The result is improved overall quality in and between both products and processes, including the getting-to-market leg of the business. Simply put, when people see their impact on the back end, they want to be more involved in the front end. With both ends of the process exposed, workers will want to contribute to building an end-user-focused process called Collaborative Process Improvement (CPI). As Cosby eluded to in his book *Quality Without Tears,* the main goal of the staff members is not to reduce the cost of nonconformance, but to delight the customer with the end product and make that customer come back for more. To connect the two worlds with the inherent idea that improvement to quality improves actual prosperity brings the consumer's world closer to the process as well. In building processes through work knowledge clusters, the CPI method becomes an inheritance for the next generation of workers in the company.

We trust that as you make your way through this book, you will find ways to make every employee a process participant. In doing so, you will educate your employees in how to be quality aware in everything they do. This awareness should not only involve product development processes, but should also extend to business processes, because they are inextricably related. In building in this awareness, you will slip in quality like a freshly baked pie from your grandmother at the back door—everyone will welcome it because it comes from a family member.

How to Read This Book

This book is intended to be referred to often. The margins are wide so that you can make notations. We expect that this book will become a guideline that you use and refer to often. It is our expectation that you will turn down corners of pages, write in margins, make copies of checklists, and utilize the information in this book to the fullest.

The book is written to prompt you to integrate the chapter aids and exercises into the fabric of your own processes. You are encouraged to use, modify, and update the example templates found here and to mark up this book as you chart your own journey to understand your work world.

We describe our prior work experiences in the boxes that appear throughout the book. Since we do want you to refer back to this book often, we have made key areas easy to discover. The following graphical icons* will help the reader to identify key areas of the book. Use these to easily reference key learning and help files. The icons will help you identify which key that particular portion of the book is defining. You can also make up your own icons to help you navigate. Don't be afraid to write in this book, and don't be afraid of your journey across the landscape of your own company's world.

Collaborative Moment—realization of collaborative benefits

Shining Moment—breakthrough; a notable event

*Graphics by Rodney Kadura.

Dollar/Time Savings—activity that impacted schedule, return on investment, and so on

Global Application—an issue/task that had global impact/results

Aids—templates, tips, tools, checklist, and things you can use

Exercise—team exercise, team building, and group activities

Hazards or Risks—lessons learned, pitfalls, warnings, things to avoid

Introduction

OUR WORLD HISTORY AND THE CASE FOR CHANGE

At our medium-sized software products company, we leveraged non-Quality personnel and effectively increased participation from 0.4% of the organization to 12% in a few short months. By using the methods described in this book, you too can bring Quality out of the shadows and into the forefront of everyday work. As you will see as you read this book, an important part of this Quality approach should be in training the organization, the whole organization, about what Quality means for your products. Relying solely on an organizational Quality group for overall quality is difficult and can fail if you do not engage the individuals creating the product in the process. Quality should be the thread by which the fabric of software is built. In order to accomplish this, you need to engage people actively in all stages.

By identifying the right people to help, we saved time and money in training on several levels. We were able to train our Thought Leaders on delivering the vision of the overall process as well as specific implementations of our processes. We saved hundreds of thousands of dollars by building digital training modules and reusing the modules. This made our quality process and training easier to attain and sustain. We were able to personalize the process for local cultures along with tool applications. When we held live training sessions, we followed them up with interactive learning sessions. Although participants initially resisted these sessions, they eventually gained valuable understanding of how to apply what they learned, and, most importantly, why they needed to be trained in that area.

We bolstered our defined processes with checklists and an intranet site that provided a constant and consistent way to communicate with everyone. In this way, we were able to continuously improve beyond boundaries of Six Sigma and even reporting structures. We engaged diverse user groups, which allowed us to communicate more effectively, and ultimately eliminated finger pointing between groups.

The organization as a whole engaged in using and understanding standard checklists that we consistently improved to reflect the needs of our groups. We then tied these checklists to metrics that we could easily understand and display. In addition, we automated important improvement tasks into a project management tool with which we could view progress and track results.

We honed our communication skills by engaging personnel in lunch-hour forums each week and integrated improvement updates into regular project team meetings. We also found that assigning personnel from one project as advocates on a different project greatly reduced the fear of repercussions from reporting less than perfect results in project teams. We did this with less than 2% of the entire group's time collectively.

We engaged our thought leaders and management in promoting the process improvements. We found that we had a 1000% increase in participation in CPI when we used our Champions (thought leaders) as recruiters for the program. Including management at all levels allowed us to be able to integrate process improvement activities into yearly performance reviews, and essentially eliminated process apathy.

Despite traditional advice that you should not share metrics across teams, we found that sharing them created a sense of urgency and friendly competition.

Overall, the programs that scored the highest on our scoreboard were also the programs that were delivering quality products on time. We started with only one group in full compliance with our guidelines and progressed to over 80% of our projects in full compliance in just 12 weeks.

With an investment of only 20% of an advocate's time, we were able to build a culture of improvement flow based on optimizing how we all worked together, which encouraged everyone to be accountable for quality. This resulted in a final payoff much larger than could be measured empirically. We learned that visualizing the end product in the user's hands and veering away from the theoretical product was an inheritance we could draw on for future products. This became an inheritance to future generations of our company whom we had not yet taught. We also learned that an investment of an advocate's time in overcoming not only the cost of a product's nonconformance, but also customer indifference, could be used to build a culture of product improvement based on optimizing how we all worked together. It resulted in job satisfaction, customer satisfaction, and business success. We were able to not only do more with less, but also work harder and smarter.

Special Thanks and Acknowledgments

To the Lord above, for teaching me that adversity leads to opportunity and for providing wonderful people to work with, including many who have collaborated with me in building effective processes. To my husband and ever patient family who taught me that without the support and contributions from people who know their own worlds, the whole world cannot be accurately represented.—C.L.Y.

To my parents, grandparents, family, and friends, for the collaborative support they have given my book and my life. To my wife, for teaching me that things are not always as they seem and that even when I look in a mirror, I am staring at an inverse of reality. To my daughter, my world is at your feet.—J.D.F.

We would like to thank the reviewers who provided useful suggestions for improving this book and also to gratefully acknowledge Cinda Cyrus and Bryan R. Henderson for thier editorial assistance in preparing the foundation of this book.

CHAPTER 1

Your World—Understanding Your Situation and Preparing First Steps

For those who have seen the Earth from space, and for the hundreds and perhaps thousands more who will, the experience most certainly changes your perspective. The things that we share in our world are far more valuable than those which divide us.

—Donald Williams

1.1 THE SITUATION

Ptolemy, a Greek mathematician, was also a mapmaker. His early maps demonstrated what happens in a lot of companies. In these maps, the world that Ptolemy was most familiar with appeared larger than the rest of the world that he knew about at the time. In fact, there were parts of the world that were actually missing, in addition to some mysterious "faces of the wind" that blew on the world and created havoc, as Ptolemy knew it. Isn't that how it is in business today? The worker buries himself in his area of work and hardly looks up to see the big picture, whether the worker knows it is out there or not. Then some strange wind blows in and the world gets churned into chaos. Later, the wind stops blowing. Now the environment is completely changed and only those that can quickly adapt to the new environment will survive.

Imagine then that you have several different Ptolemys drawing maps of your organization. You know the software world very well and your map of the software world is very big. The winds of

Collaborative Process Improvement. By C. L. Yeakley and J. D. Fiebrich
Copyright © 2007 IEEE Computer Society.

change blow in, but the average workers do not understand or want change. They just want to code, and they want everyone else to get out of their way. Their greatest wish is that the wind will stop blowing. Then there is the marketing organization that sees customer requirements as being very large and the software world as something very small. This myopic view is replicated throughout the organization, from engineering, to finance, right up to the Chief Executive Office.

Now imagine taking all these maps and overlapping them. Take all the oversized features and hook them all together to form a single map. The world then comes back into proportion and the picture is complete. Things such as company goals, visions, and bottom-line revenue requirements have replaced the winds. The continents become mechanisms for delivering products.

Look around your company. How many personnel are dedicated to improving the quality of your product by keeping an eye on the process that creates it? Is quality improvement relegated to a select few who have knowledge of quality principles? Is there general miscommunication and misunderstanding of company goals and visions? Does there appear to be a rift between what management says and how the employees execute their everyday work?

Are there many different worlds, or perspectives in your company? It is hard to ensure that everyone shares the same company (world) view because you have only one set of eyes and ears and can only see from your own perspective? No matter how good you are, you do not live every day in each employee's shoes. Even if you bootstrapped up from mailroom clerk, to the factory floor, to management, and then to executive management, you would only know how it was *then,* when you were in the moment. Every day your company evolves, whether or not you direct that evolution. It can go very well, or it can go poorly.

As Ptolemy saw the world as being larger and more detailed where he was, and the rest of the world as smaller (or nonexistent) where he could not be, so do employees see their own areas in greater focus. But what is wrong with that picture, really? Ptolemy's view of the world also had "strange forces" depicted as faces blowing wind around in his world. Is Ptolemy's view the way you see executive management (or your Quality Department) in your organization?

The person living the process can improve it best if he or she can collaborate with others who "feel their pain." As long as the improvement can be later scoped back into the balanced picture, there is nothing wrong with the Ptolemy view of the world. You can use Ptolemy's principle to your advantage! You just need to be careful to ensure that you can fit your world back into the big picture so that you can still deliver what the customer wants. Your challenge is to paint the world as the customer sees it, and have your employees be able to paint their world in a light that the customer ultimately finds pleasing.

How many times does your organization deliver what customers ask for or what they truly want? Is the end goal to deliver on time and within budget, and are your processes geared toward effective delivery of product without regard to how your own employees perceive the process? J. Davidson Frame, author of *Managing Projects In Organizations,* says that even though a product may be delivered on time and within budget (the rally cry of many program management offices), "If the final deliverable does not really address the customer's needs, or it meets with customer resistance, the customer will not employ it" [2]. We boldly declare that more management is not going to prevent customers from being dissatisfied in the future, nor is the problem going to solely be solved by your Quality Department.

1.2   PREPARATION

What we propose is that you take a good look at your organization and ask if it is set up for success at the individual level. Are there enough fluid processes in place so that your team can collaborate on the overall delivery process? Are you not only enabling your team to work smart, but are you building that into everyday work so that you do not have to launch the next Six Sigma effort (as did GE and Motorola) or business process improvement (as did Dell) effort to solve the problem?

Is your organization a nimble seagoing canoe or a hierarchical raft on which crews just go with the flow to get by? [3] William R. Daniels and John G. Mathers, authors of *Change-ABLE Organization—Key Management Practices for Speed and Flexibility,* describe the seagoing canoe as having members who sit in their own seats and keep their eyes on their own goals while they all strive to reach a common goal. Each person is in control of his or her own area, with the leader out in front with the ability to see just a little bit of the larger picture shortly before the entire team does. Each paddler is responsible for making his or her own subtle adjustments, and they wind up encouraging each other to do better, to keep up, and to keep a careful watch on the condition of their part of the boat. We feel that the real root of delivering a quality product to a customer is enabled by the ability to build a collaborative process improvement (CPI) environment that does not focus solely on large, bureaucratic improvement processes, but instead uses those to validate some of the stand-alone projects under the working team's care.

The solution we propose will rely on the collective team to row in the right direction together while they begin to treat each other as part of the family. Remember that what we are trying to do is "sneak quality in" to every part of your organization by giving everyone a role as a family member, complete with all the rights that come with being a family. But our family is looking for something special. They are looking to build a CPI effort that is sustainable and excites them, because they are part of the ongoing success of building a Quality process.

The truth is that no one wants to belong to a losing team—we all want to be winners. The average worker is driven to deliver according to his or her commitments and rarely has time to think about process improvements or the effect of the process on the quality of the product. The advantage of building an organization that understands how to collaborate for process improvement is that the improvement process is built into everyday work. In this way, your teams can continually improve products and processes while continuing to deliver products.

Before you attempt to move your organization to collaborative improvements, you will need to take inventory of what your organization looks like in terms of ability and infrastructure for change, and exactly where the oft-quoted but never quite attainable "low-hanging fruit" is located. When looking for areas of opportunity, also be careful to note any pockets of excellence. It is from those pockets of excellence that you are likely to find your thought leaders and process advocates. As Malcolm Gladwell points out in his book, *Tipping Point*, you need to recruit a thought leader—someone who is known and respected among the non-Quality community—to help speak those Quality words for you and manage by influence. [4] This then, will be a good place to start recruiting help.

Once you understand the big picture of how your organization operates, engage with all levels of management to ensure that your teams will be enabled to make changes. Part of communicating the direction is providing the picture of the mission on the horizon. Target the overall goal, so that the nimble team you are about to create can hold that goal in mind while they are busy rowing the boat.

The questions you need to answer as you start your journey to building a CPI environment are:

- What is the vision and mission for your organization?
- Where are the likely first places to improve?
- Is the management team dedicated to supporting and enabling collaborative teams?

- What is the first big win you will attempt?
- Where are you going to find people to help in this effort?

1.3 VISION

To answer the vision and mission question for your organization, you should meet with several coworkers and read the vision and mission aloud. Then go around the room and ask each person what it means to them. You may be surprised to find that each individual has their own translation of the vision and mission. Sometimes mission statements are so expansive that they fail to focus on reality. Other times, statements are so tight that they have no meaning, as discussed in Jim Collins book, *Good to Great: Why Some Companies Make the Leap . . . and Others Don't* [5]. Or perhaps you all understand the vision and mission of your organization, and you have it mapped down to the task, as is done in a cascading goals method.

What you really want to know is in what direction is your company going and do you have a way to build CPI into your world. You may have to be creative to insert CPI wording into the translation of the vision and mission, but you are sure to lose out if you do not build some specific references that validate your ability to charter a CPI effort.

1.4 IMPROVEMENT AREAS

Once you have had the meeting to understand the vision and mission of your company, you will want to decide on what areas to focus on. What CPI team would you like to start? In reality, you want to create an organization that is a continuous CPI world, where process improvements are the way of great business, and you do not have to launch individual process improvement projects unless there is a specific, single targeted improvement goal. You want to build an organization that holistically looks at process from the worker level and engages with them to improve the organization in an ongoing manner, without the need for workers to be chartered and prove their worth—sort of the Mobius loop of process improvements, with no beginning and no end. At least, that is how you want the organization to see CPI.

In reality, you will have to start somewhere—to have your own "big bang" and bring life to the CPI effort where no life existed before. To do this, you will have to pick a framework or an area to launch. It is a well-known fact that the pain of change has to be less than the pain of staying the same. So look for the pain! Where could a change in process relieve a lot of pain? In this book, we chose software development as our main pain point. Our organization selected the Key Process Areas of the Software Engineering Institute as our framework. The model actually is irrelevant as you are building up a collaborative process that should eventually take on its own framework and build its own world view. Possible pain points are manufacturing, development, product delivery, design, and communication.

Pick an area in which you can find a standard to follow and then tackle it by engaging those involved in doing the work in those areas. Oftentimes, it is very easy to figure out where the pain points are. These are the things that people feel helpless about or talk about a lot. They are the things that people say "Welcome to the Company" about—the things that your organization is well known for—and about which employees often may voice grumbling, hopeless comments.

1.5 MANAGEMENT SUPPORT

Assuming that you have found your niche inside of your company's vision and mission statement, and you have identified your current pain points, you will need to garner management support. You will need to engage with management at varying levels to get agreement that you are on the right track. As in the identification of pain points, select the appropriate management chain to champion your effort and begin to build an alliance with a few key managers. These managers will prove to be your CPI sponsors of the future. In the beginning of the effort, you are just trying to get to critical mass so that your "big bang" won't fizzle before launch.

Management will have a large role in the CPI efforts that you will sustain throughout your company's lifetime. The first step, however, will simply be to get them to understand how your CPI efforts will net the company big gains. Your management should be able to direct you in ways to communicate with varying levels of upper management so that a general awareness of the beginning of the effort will be in place before you engage with the workers. In this way, you will build in credibility for the overall effort. Further discussion of management sponsorship will be provided later in this book.

For a starting place, you will need to get a few managers to understand the big picture of process improvement. You might even start with those managers who are big supporters of the traditional Six Sigma or business process improvement efforts. Tell them that this process differs from those in that it engages people at all levels. Explain that it is not just a special process improvement effort but a continuous, built-in effort. You may have to pull out a few figures for them and, hopefully, you will find enough in the pages here to build your own case for world change.

1.6   YOUR FIRST BIG WIN

Once you have your company's vision and mission well understood by a small starter team and have identified the main pain points in your company and some likely candidates to launch your first CPI effort, you will still need to secure some management support before you are ready to determine what pain point, when relieved, will get noticed. Those pain points will be key in your success because it will be your job to relieve the pain with an improved process. It is very important to the success of change acceleration that you can demonstrate some wins very early on in your efforts.

We selected the Key Process Areas for Software as the overall framework and the formal review process as our "relieve the pain fast" solution. We knew from prior experience and also industry literature that if we could train our organization to perform reviews properly (and with the proper participants), we would be able to prove that our CPI effort would have immediate benefits. We knew we needed to get people excited about the overall process but also to show that their ideas and inputs could result in big wins. These wins were not only wins for the company, but also relieved personal pain on the job, resulting in personal wins. We were driven to show the workers what was in it for them.

Many of the leaders in your organization probably have projects that they know would succeed if they could only get momentum behind them. In software, the formal review process is generally a big win. It won't be sustainable outside of an overall process, so it is a very good big-win selection.

If you are in a different industry, you are likely to encounter a similar situation. For International Organization of Standardization (ISO) requirements, which are prescriptive rather than being a framework, perhaps the big win would be the document management processes. You can change your mind at any time during this process and look at a different framework on which to focus. Or you may already have a product delivery process that you can use as a starting place to build off of. Either way, the important idea here is to get your team together and listen to what they tell you. This brings us to our last initiation area, the people.

1.7   WHO WILL LEAD?

The answer to who will lead your effort lies within your organization. Depending on the nature of your changes, you will need to select appropriate leaders. These people may not see very far ahead of the rest of the team, as in the seagoing canoe example we discussed earlier. In our software case study, the leader was the Business Process Director. In your organization, it could be the Operations Director, the Quality Director, or the Finance Director. It depends on the structure and size of your organization. Whoever that person is, they should be well experienced in change management and preferably in a technical area that your company uses to build products. Once you have determined who the leader of this effort is, you will be ready to start your CPI effort.

CHAPTER 1 SUMMARY

This chapter introduces the first things you should address when beginning to implement a CPI program in your company. Key ideas presented were to assess the following:

The situation—Ensure that you understand the different perspectives in your company. Begin to build the map of your organization.

Preparation—Determine your infrastructure for organizational change. Identify areas of opportunity that fit into your organizational vision and mission. Enlist the help of management to enable collaborative teams.

Vision—Ensure that you see the place of CPI in your overall corporate goals. Identify where these efforts can contribute to the vision and mission of your group.

Improvement areas—What framework could work for you? Will Software Engineering Institute models fit? Do you need to combine any framework with International Organization of Standardization requirements? Where are the main points that affect your products?

Management support—Communicate with management and find secure upper management sponsorship.

Your first big win—Choose a process that will yield an immediate, recognizable win for the organization, one that the workers will recognize and is tangible.

Who will lead?—Select a central contact person who will be the cornerstone for all CPI efforts.

CHAPTER 1 AIDS—SAMPLE INITIAL BUSINESS STATEMENT

Use the following example business statement to begin to build your own. You should decide on five to seven items in this list. Do not spend a lot of time worrying about style and format; just ensure that your statements are reasonable and agreeable to your team. This will serve as a starting point for moving forward.

1. Acknowledge that software plays an ever-increasing and vital role in company's product sales. Establish and agree on software quality objectives and have the ability to measure ongoing status.
2. Determine whether or not quality objectives are consistent and in alignment with day-to-day operational objectives. That is, when faced with quality versus time-to-market trade-offs, establish criteria on which to base these decisions.
3. Determine a business model that relates to software. Ensure that it is consistent across all the business units and organizations: development, release, and support.
4. Understand the budgetary implications and support mechanisms that must be in place for the business model (software-focused activities).
5. Create a work environment that is technically stimulating and furthers the career goals of employees.
6. An increased focus on software will lead to an overall improvement in product quality and, hence, revenue goals.

CHAPTER 1 AIDS—SAMPLE CASCADING GOALS

This is an example of how to cascade goals from a strategy down through goals and objectives. This should be a natural progression from your business objectives in the previous example. This does not have to contain elegant wording or highly polished formatting. Some

organizations put this in an outline form in an Excel spreadsheet for easy tracking. Afterward, they track each goal and objective at a high level with a red/green/yellow stoplight symbol for at-a-glance reviewing. Managers can track progress of their teams and hold them accountable in their performance plans. Objectives can be further broken down into tasks for increased tracking at the individual level. These tasks can also be integrated into the performance plans of individuals.

Business Strategy—Quality Through Core Processes

1. **Goal**—Drive software quality awareness to all levels of the software development organization and corporate management. This will result in the ability of all levels of the organization at all phases of development to answer the question, "What is our software quality?"
    1.1. **Objective**—Implement a company-wide software (and software process) metrics program to document software quality levels and actual and planned schedule and cost tracking in order to support business decision making.
2. **Goal**—Evolve the software development process to achieve a defined and repeatable process (using Software Quality Engineering methods) across all development teams, product organizations, and business units.
    2.1. **Objective**—Evaluate the state of the art of software development, and train the software development organization on current industry best practices and methods.

Business Strategy—Product Leadership: Delivering Products/Reducing Cycle Times

1. **Goal**—Create a software development culture in which the expectations of chosen target markets are fully understood.
    1.1. **Objective**—Determine effective ways to define customer requirements/expectations.
    1.2. **Objective**—Adopt a product life cycle management model that facilitates proactive analysis and decision making.
    1.3. **Objective**—Reduce the average elapsed time required to move new software products or releases from early adopter to full-scale volume deployment by our customers.

CHAPTER 1 AIDS—SAMPLE SPONSOR SELECTION WORKSHEET

Use this checklist to evaluate possible sponsors for your change efforts.

1. Does the potential sponsor have enough influence in the company to increase visibility of the efforts?
2. Can the potential sponsor garner support from senior management?
3. Is the potential sponsor willing to take a highly visible role in the effort?
4. Will the potential sponsor be personally accountable for results of the change efforts?
5. Will the potential sponsor follow through with commitments?
6. Will the potential sponsor ensure that efforts will be tied into company strategy?
7. Does the potential sponsor understand the need for change and will he or she communicate changes widely?
8. Is the potential sponsor at a high enough level in the organization to influence change and ensure that employees will be held accountable?
9. Does the potential sponsor believe in change initiatives and will he or she make CPIs a high priority?
10. Will the potential sponsor take an active role in change teams and be an advocate for supplying needed resources?

Chapter 2

Welcome to the World— Establishing Advocates and Champions

There are many qualities that make a great leader. But having strong beliefs, being able to stick with them through popular and unpopular times, is the most important characteristic of a great leader.

—Rudy Giuliani

2.1 RECRUITING CHAMPIONS AND ADVOCATES

By recruiting select personnel—those doing the development and support-services work—you can build a team that understands the mechanisms of change and the need to engage the organization. These individuals are called Champions. Champions are the main points of contact in key process areas such as configuration and requirements management. For whatever reason, be it qualifications or the natural inclination of the people to gravitate toward them, they are the people in the organization who are respected as the thought leaders and natural drivers, and, because of people's affinity for them, can rally support, because people trust their opinions.

The other people to be identified as playing a necessary role are called advocates. An advocate is a designated person who lives day to day with projects and forms the very fabric of project teams. This person's duties directly relate to tracking and reporting on project progress from a process perspective. You must train both the champions and the advocates on both Quality principles and soft skills so that they can effectively communicate with the people they work with daily.

Once you select your champions and advocates, you must make them very visible in the organization. In addition, they should produce a set of results that management can expect. In the article "On the Needs of Selves and Societies," Michael Kearl, Professor and Chair in the Department of Sociology and Anthropology at Trinity University, stresses that it has been shown that people in groups choose to behave a certain way not because of what they consciously think or because of their experiences, but because they are reacting to perceived expectations [6]. You can encourage the right expectations from your team by holding up champions and advocates as models of what management expects from the organization. They must back this up with actions—and attention. This will help the employee see that this is expected behavior and that this is "standard operating procedures" in the company.

Recruit individuals from all levels within the company, and be extremely selective in doing so. Look for people with seniority, superior technical talent, and human relations skills. The first pool of people that management may offer to you may be individuals that are not performing or are slated for termination. Do not recruit or accept those who are on probation, are mediocre performers, or whom the business does not respect [7]. Testers, developers, and program managers all enter the arena of Quality at the same level. Although the individuals remain in their current reporting structure in the business, they should also report to the Software Quality Engineer, as shown by the dotted lines in Figure 2.1.

This dotted-line reporting structure must be acknowledged and accepted by both the business and the Quality Department. You should staff your Software Quality Engineer position with an experienced individual with an engineering degree and at least 10 years

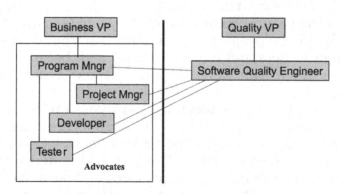

Figure 2.1. Advocate reporting structure.

of experience. His or her main focus will be software process improvement at the company level.

Over a period of twelve months, although the company decreased its employees by 11%, the ratio of personnel performing quality assurance tasks increased from 1:250 to 3:25, as shown in Table 2.1. The increased presence of individuals performing assessments, evaluations, reviews, and meeting facilitation created a sense that the company was making a commitment to the quality of their products.

Once these champions and advocates are recruited, they must be trained to perform their new tasks. Although few of them will have ever participated in Quality activities, they soon will be not only performing these tasks, but will be advocating Quality processes.

Recruiting Hindsight. We have since learned that it is extremely valuable to have all new hires participate in the Advocate Program for the first 6 months of their employment. This program gets them the immediate training they need regarding the company's processes and procedures. They get great exposure via projects and other advocates. But more importantly, the new hires bring with them the things that worked at their previous employment. College graduates oftentimes bring the latest techniques from academia, leading to more continuous improvement!

The importance of the advocate role cannot be minimized. If it is handled well, the role will be seen as essential to the organization. If part of the new hire on-boarding process is to participate in

Table 2.1. Quality versus nonquality personnel distribution

Personnel	Prior to Project Initiation	At Project Initiation	After First Year
Non-Quality	250	250	225
Quality	1	2	2
Advocates	0	0	12
Champions	0	0	13

the Advocate Program, you must ensure that this role carries great weight. You might consider using the advocate role as a condition for continued employment. In this way, the role will not be minimized because it is considered something only the new hire has to live through. It will be instead seen as essential and needed.

This, in turn, will take advantage of a simple human trait: To be needed is to have value. "If people depend on me, then I am important." It has become popular for corporations to hire relationship consultants to help define the process of human interaction on the job. This practice has shown mild success. The difficult part is that people still paint relationships in terms of roles. In the book, *The Moral Animal: Why We Are the Way We Are; The New Science of Evolutionary Psychology,* Robert Wright explains that when employees interact according to defined roles, the level of interest depends upon the status of the role. This causes people to give the most attention to those above them [8]. When you are in the business of CPI, you have to figure out every way you can build a team that is motivated to achieve. Making the role to be perceived as a much-needed role will build in a sense of pride and ownership that will motivate not only your selected champions and advocates, but will also have others waiting in line to participate in the program. Once you have provided a means for individuals to be motivated, you can worry about the individual skills they need to be successful. One of these skills is the ability to organize the information they will be gathering.

2.2 ORGANIZATIONAL SKILLS

Lack of organizational skills can create more chaos than the chaos you were trying to avoid in the first place! Signs of organizational issues include not being able to find process assessmentassessment information easily, not informing the team of results in a timely manner, and conducting assessments only sporadically. Of course, the largest sign of an organizational issue is lack of adoption of the principles of your process improvement program. Your goal should be to get your process improvement program to fit into the organization. In order to do that, you will need to provide aids in getting that accomplished.

Remember that while you are engaging non-Quality personnel in building lasting process improvements that will lead to a level of Quality that will delight your customers, you will still need to engage with a few software quality engineers who have expertise in that area [9]. You will need to figure out a way to spread their knowledge deeply into the organization through your advocates. Many times, the people that you have identified as good communicators are not the same people who have good organizational skills and are naturals at building lists and following them.

A good way to give your advocates a way to organize their information is to provide aids. Particularly valuable organizational helpers are checklistschecklists and activities added into your project management schedules. By providing appropriate checklists, you make the work easy and help in organization information; you do not have to rely on individuals to set up their own reporting scheme.

By building advocate activities directly into your project-management scheduling tool, both you and the advocate can track when to do an assessment and what processes need to be assessed. You should utilize the repository activities.

Provide an intranet page for posting information, including templates and process helpers. In addition, each team should have storage space on your internal document-management system from which all personnel can easily retrieve project information [10].

The point being made here is that you need to provide some basic tools that make it easy to organize the information that the advocates are using to assess a project's health. The organization should review, accept, and own the tools. These become standards for reporting and organizing process information. Once your Advocates understand how to arrange the information physically, you can move on to how they arrange the information in their minds.

2.3 THE ASSESSMENT MINDSET

Advocates might demonstrate more assessment personalities than you really want! An advocate writing up projects on minor noncompliance issues might demonstrate evidence of an assessment problem or be dogmatic about your organization's chosen processes. Alternatively, an advocate can try to be the "good cop" by forgiving blatant noncompliance. Another symptom of the lack of an appropriate assessment mindset is that your advocate is simply unsure about the sequence of events surrounding an assessment. This can be followed by a general reluctance to accept a process role.

At our company, we addressed these problems in the following ways. We provided extensive checklists and activity descriptions. We held monthly Advocate meetings. In these meetings, all advocates joined Quality personnel in a general discussion. This forum provided a free-flowing question and answer session as well as training opportunities. By doing this, we allowed free play and creative thinking.

By providing the checklists and activity descriptions, you have a baseline that the advocate can reference. This provides a basis on which to have discussions at the project level and above. In this way, you build in the sorts of questions that should be asked of a project team, and help drive an Advocate's style and eventually, his or her ability to successfully conduct an assessment [11]. By reviewing the documentation from the weekly reports, you can see when an advocate strayed from the guidelines.

Forums provide avenues to discuss processes in a nonthreatening way. Advocates feel free to bring up what they think are stupid questions and all are welcome to voice their doubts and concerns. In addition, you may have guests attend who have a particular area of concern about which they are passionate. Forums provide a way not only for advocates to learn and communicate, but to adjust their attitudes about their roles by hearing what the organization is saying us about the overall process.

What could be a flawed process becomes an effective and functional process due to the standardization of how advocates across the organization approach their project teams and the extensive communication encouraged while building those processes. By having this weekly meeting of the minds, in the spirit of improving the process, the advocates can learn how to approach their teams in a nonthreatening way. In addition, it makes it easier to communicate the visionvision of the overall process.

2.4 PUBLIC RELATIONS, PROFESSIONALISM, AND COMMUNICATION SKILLS

Of course, it is easy to train by concentrating on the specifics of a particular methodology and guideline, compared to trying to train behaviors. The personalities and communication skills of advocates are certainly as diverse as those found in the engineering populations with whom they work. Not every person is adept at tactfulness, professionalism, or managing a project team. These are the very skills essential to communicating process issues and getting results. You can tell whether your advocate is being effective by the types of comments you get postreview and at milestones. If you hear things like, "I'm afraid of our next assessment," or, "I hide every time I see the advocate coming," then you know that the advocate interface is not effective.

You can mitigate these sorts of problems by coaching and training. By modeling dialogues for advocates, you can teach them how to communicate both good and bad news. It is important to be able to deliver constructive feedback concerning process adherence. Several good books offer excellent information on communication and accountability. One such book is *The Oz Principle* by Hickman Connors, and Smith [12]. It has excellent guidelines on how to get individuals to take responsibility for their own work and to seek feedback on how others perceive them.

You could consider this as a way of providing constructive feedback principles to a project team rather than to an individual, but it reaches deeper than that. Constructive feedback delves into how an organization looks at its work products and the means to get there. One key to driving effective process improvements is the ability to deliver a critique in a nonthreatening way while getting the point across and building a community that wants to listen.

Some questions you can ask your organization are:

- How well do we listen as an organization?
- How tuned in is our management team? Do they understand the value of feedback?
- What is our communication style?

By understanding how an organization communicates, you can coach the team in the methods they use to communicate. The advocate will need to fit into the culture and drive for change in a way that the organization will accept.

Here are a few examples that you might consider as you decide how your organization communicates. A hierarchical organization that is top-down driven and highly structured has a different communication style than a matrix organization. There may be an open-door policy, but few use it. Communication channels here are different from those in an organization that thrives on escalations. In the first case, the advocates would write formal reports that they would deliver to upper management after they delivered them to the project team. The parent of the group would then read the report, monitor, and control the team. In a matrix organization, teams expected teams to do their work with little supervision. They only report extreme cases of noncompliance via an escalation mechanism. The ability to escalate and the motivation to do so are tightly integrated. Empowerment seems to be relegated to the "old school" method of Quality when, in reality, it has everything to do with the ability to successfully monitor and report on Quality process activities.

2.5 EMPOWERMENT

Without motivation, empowermentempowerment is meaningless. But without empowerment, motivation to improve is an unlikely outcome. Although Quality standards have been mentioning empowerment of the individual for a long time, the actual tactical execution of empowerment is tricky and hard to institutionalize-institutionalize. In addition, companies tend to thin out overly heavy Quality processes and organizations in difficult times. After all, they are not directly contributing to creating the product, so they are the first to go in lean times. This undermines the importance and the perceived expectations of the Quality group as a whole. A careful review of the evolution of Quality standards reveals the disturbing but common idea that Quality is something that is imposed on a project rather than built into it; that individual empowerment is less important than an overall Quality program.

The U.S. Department of Defense used Mil-Q-9858A as a quality management standard for more than 50 years, ending its unique authority in 1994 in favor of ISO 9000. However, some things were lost in the translation. Mil-Q-9858A had a strong *employee empowerment endorsement.* It said, "Personnel who perform quality functions shall have sufficient, well-defined responsibility, authority and organizational freedom to identify and evaluate quality problems and to initiate, recommend or provide solutions." In government lingo, "shall have" is a directive, so the statement is a declaration of empowerment.

Unfortunately, ISO 9001:1994 is less forceful in its support of employee responsibility. The words sound the same, but on close reading we find a company may "define and document" the authority of an employee who "needs the organizational freedom to (initiate, recommend or provide solutions. . .) [13]." ISO 9001:2000 drops the issue altogether [14].

In their book titled *Management of Organizational Behavior*, Paul Hersey and Kenneth Blanchard described the modes of employee empowerment, tracing them historically [15]. At the low end, they list the "tell" mode, in which management literally tells the employee what to do at each stage of the process. This mode worked reasonably well in the early part of the 19th century, with a workforce of illiterate and inarticulate immigrants.

The "delegate" mode is the high end of empowermentempowerment, in which the company delegates employee responsibility, and, in turn, the employee is responsible for quality performance.

Although most companies boast of their relationships with employees, the reality may be quite different. Empowerment is an ethical issue because it establishes the authority of employees in the performance of their effort. It therefore affects the quality of human decisions.

An essential part of the CPI philosophy is to ensure that your improvement teams are empowered on many different levels. In order to ensure that they have the traits and full understanding of the improvement process, you must consider a few areas in which they might need some training.

2.6 INITIAL CONCERNS

At the initiation of an Advocate Program, the first question that all parties involved must ask will be "How much of my time will this require?" People will want to understand how much weight they will be carrying in addition to their "real" work. There will be an initial investment for sure but, as time goes on, this will become second nature to your group and it will become just the way you do business. You can estimate how long it would take in the beginning, but this will only be a target.

The interesting phenomenon that you will observe is that once people engage in defining the improvements, they start investing more of their time in them—voluntarily! They want the change to stick, they feel the results of the improvement, and they want to continue.

Of course, if you have a quality or business organization, you will be able to utilize these people to help in building checklists and facilitating improvement sessions. Depending on how many people you have available for this sort of work, you will be able to skim some required time from your advocates and champions. If you have a very small group of one or two, then you will need to engage your Advocates and Champions at a higher level.

Your Business Process Group can provide many things such as building training components, training, Web page creation and maintenance, and direct mentoring of new processes. The role of your Quality or Business Process Director will be to see the big picture for the purpose of facilitating and coordinating work done at the individual level. Remember the Ptolemy map in the first chapter? The job of the Director will be to pull all the different perspectives together into a coherent whole. It will also be the job of the leader to report metrics and to communicate at all levels of management. It is actually a large job that could require some assistance depending on the size of the organization and how many (if any) other process personnel are available.

Initially, we had no idea, but started with a target of 10 to 20%. A year into the program, we estimated the hours the advocates spent. At first glance, it appeared that individuals spent 40% of their time away from normal tasks. This was alarming. After diving deeper into the data, we determined that the advocates were spending 60% of their time doing engineering and 20% of their time performing code and documentation review, for which they were responsible anyway!

Only 20% of their time was spent specifically on advocate activities—configuration management assessments, quality assurance assessments, and process assessments, as shown in Figure 2.2. This division of time is an amount with which all parties should be comfortable.

The excitement over winning and making incremental opportunities for new wins will help your organization be self-improving. After a while of doing this, your workers will wonder why this was never done before. Your challenge is to make your CPI conversion look and feel easy to accomplish so that everyone wants to pick up a little of the weight. After all, "Many hands make for light work."

CHAPTER 2 SUMMARY

This purpose of this chapter was to review methods for establishing champions and advocates. Each section is summarized below.

Recruiting Champions and Advocates. Champions are main points of contact for key areas, whereas advocates are the workers who are engaged in day-to-day project activities. Both must

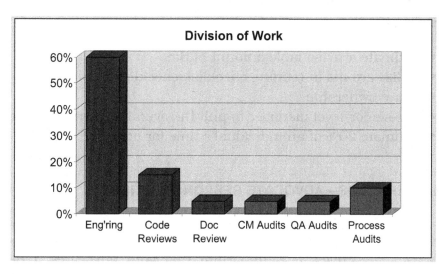

Figure 2.2. Division of work.

have high visibility in the organization and be given essential roles.

Organizational Skills. Advocates should spread their knowledge deeply throughout the organization. Checklists and specific tasks recorded in project management tracking applications will assist in driving the right activities into day-to-day work. Basic tools such as process intranet sites and data repositories will assist in providing an organizational infrastructure for sustained change.

The Assessment Mindset. Advocates should baseline their activities by using the defined checklists and activity descriptions. Engaging all advocates in a monthly forum will ensure that they can continue to learn and share information. This will self-adjust attitudes about their roles to ensure standardization of approaches and to extend communication.

Public Relations, Professionalism, and Communication Skills. Advocates might need help in determining how to deliver both good and bad news. This can be achieved by modeling dialogues, coaching, and understanding how the organization as a whole communicates in order to fit the message to the terms the current culture can identify with.

Empowerment. Improvement teams, from workers establishing an improved process to the advocates who help monitor the process, to the champions who help drive the processes, must all feel as though they are valued and empowered to make a difference. This empowermentempowerment must encourage workers to be able to feel that they can indeed change their world.

Weight Lifting

- Estimate a reasonable amount of time
- Utilize groups in refining checklistschecklists
- Share ownership
- Use senior-level manager to pull the overall picture together
- Estimate 20% of an individual's time for advocate activities

CHAPTER 2 AIDS—ADVOCATE RESPONSIBILITIES

 This list of responsibilities can be adapted for your organization according to your goals and objectives. It is important that your advocates understand their roles, as this will help them to perform all the duties required of the position. It will also establish expectations for the entire organization and, in particular, the groups with whom they are working.

- Remind teammates of process. Be the conscience of the team.
- Monitor the project to ensure that teammates are following processes, policies, and standards.
- Be aware of project plans; ensure that all planned elements exist.
- Monitor test results (this might take place at the end of the project).
- Report to the project group on how well processes are working/being complied.
- Report any issues or improvements needed to the Process Director at least monthly (include quality assurance risks).
- Go through the checklist at phase transitions on the project.

CHAPTER 2 AIDS—SAMPLE CHECKLISTS

The following is an example of one of the checklistschecklists we used. In this case, this was for our definition phase of development. The advocates used this checklist to perform their duties in a very consistent way. This further defined the expectations for both the advocates and the teams with whom they were working. Notice that the checklist is very simple, with yes/no choices and a column for objective evidence. Objective evidence is just a formal way of saying "this is the proof that this task was done." Usually, this is a document of some sort or even an e-mail notification. The objective evidence column could easily be set up by the assessor to contain a link to a particular document. For more advanced organizations with project management tracking, this list could be converted by a project manager to a work breakdown structure with completion noted as well as links to evidence. Notice that the checklist even has advocate activities listed.

Definition Phase

Yes	No	Review Item	Objective Evidence
		Have you drafted the marketing requirements?	
		Have you started the project management plan?	
		Have you set up the project planning?	
		Have you considered the software life-cycle model?	
		Have you identified and documented the risks?	
		Have you determined the management sign-off date?	
		Have you identified a quality assurance advocate?	
		Have you trained the advocate?	

The following is the planning-phase checklist used for the next phase in our example process. You could also use this checklist in a schedule review session in order to ensure that all activities have been planned.

Planning Phase

Yes	No	Review Item	Objective Evidence
		Project updated in project tracking tool?	
		Roles and Responsibilities Defined?	
		Test management plan started?	
		Project plan exists?	
		Project kick-off meeting held?	
		Configuration management plan in place?	
		Risks documented?	
		Estimates documented?	
		Project schedule includes quality assurance activities? (Assessments, phase gate reviews, post implementation review)	
		Project team meetings are held and decisions documented?	
		Change Control Board process in place and used?	
		Configuration management audit performed?	
		Document repository set up?	
		Requirements documented?	
		Requirements traceability set up?	
		Relevant post mortem reports reviewed?	

Yes	No	Review Item	Objective Evidence
		Has the advocate verified that all plans (Quality, configuration management, project) are in place and flagged any inconsistencies to the right people?	
		Training planned and conducted (if applicable)?	

Development Phase

Yes	No	Review Item	Objective Evidence
		Design reviews/walk-throughs completed?	
		Schedule actual versus planned tracked?	
		Risks updated?	
		Have documents been revised as appropriate?	
		Team meetings are held?	
		Requirements changes have been incorporated into project docs?	
		Documents can be found in the appropriate repository?	
		Phase gate review conducted?	
		Changes to project have been reviewed by the right people?	
		Code is in approved repository?	

Yes	No	Review Item	Objective Evidence
		Requirements can be traced?	
		Is the project team managing and tracing requirements?	
		Has the project advocate reviewed the status of the project and reported any red flags to the right people?	
		Quality issues are reported on monthly (via report or an e-mail update to note that Quality activities are occurring)?	
		Advocate reviews the code repository to ensure that it is being used as defined in the project configuration management plan?	

Testing Phase

Yes	No	Review Item	Objective Evidence
		The bug database is being used according to policy?	
		Test results are being logged?	
		All documents are being updated?	
		Any process or standards that are out of compliance are reported to the Key Process Area Champion (if they could not be resolved within the project)?	

Yes	No	Review Item	Objective Evidence
		Configuration management audits occur?	
		Project team meetings are held?	
		Change Control Board process is being followed?	
		Documents can be found in the appropriate repository?	
		Critical dependencies are being tracked?	
		Release readiness review is being conducted?	
		Release certificate is filled out?	
		The advocate reviews the bug report and brings any open issues to the Key Process Area Champion?	
		Phase gate reviews were conducted?	

The following is the planning-phase checklist used for the next phase in our example process. You could also use this checklist in a schedule review session in order to ensure that all activities have been planned.

CHAPTER 2 AIDS—GUIDE SHEET FOR PROJECT REVIEWS

Guidelines

Perform a review of the project at project meetings and with program management at selected, predefined intervals. Compare project management issues with program management oversight and pay particular attention to interfaces and risks that may change throughout the life cycle.

Do You Have a Current Work Breakdown Structure at the Project Level?

- Is the critical path identified and understandable?
- Are there dependencies that do not appear on your work breakdown structure?
- Is the information clear and displayed where the project team can see and track progress?
- Does the work breakdown structure reflect measurable tasks of short duration?
- Do you have project objectives clearly linked to lower-level objectives?
- Do you have clearly defined deliverables at the project level?
- Have you identified your resources in the budget?

Do You Have a Current, Credible Schedule and Budget?

- Does the work breakdown structure support the schedule?
- Have you based your schedule on realistic and quantitative estimates (where assumptions were documented)?
- Does the schedule provide time for training, vacations, sick leave, and quality assurance activities?
- Does the schedule allow for interdependencies?
- Are the next 90 days detailed? Are longer time frames less detailed?
- Can you perform to the schedule and budget (resources)?

Do Project Personnel Understand the Software Requirements?

- Do project personnel understand the system clearly?
- Are interfaces clearly understood (prototyped if necessary)?
- Is the architecture and design method traceable to system operational characteristics?
- Are development requirements explicit and testable?
- Are acceptance/delivery criteria explicitly defined?
- Has the project team agreed to their commitments?
- Are requirements traceable from concept through design and test?
- Can units of code be tracked back to requirements; that is, are we only building what is asked for?
- Are requirements managed and under change control?

Are You Tracking Current Top 10 Project Risks?

- Are risks managed throughout the project life cycle?
- Do tracked risks include cost, schedule, technical, technology, staffing, external dependencies, supportability, maintainability, organizational, and marketing issues?
- Are *all* project personnel empowered to identify risks and understand the escalation path?
- Are corrective action plans documented, communicated, and managed to closure?
- Are user requirements relatively stable and managed as a risk?
- How are risks changing over time?

Do You Understand Your Current Schedule?

- Is the schedule modified when major modifications take place?
- Have people been trained to accomplish their tasks?
- Does the project avoid extreme dependence on "heroes" or specific individuals?
- Are people working abnormal hours?
- Does the software features list get adjusted in proportion to the identified delivery date?
- Has the software been designed to maximize parallel programmer effort?

Project Management and Metrics

- Do you manage by fact? Have you identified metrics to track early indications of problems, quality of the product, effectiveness of processes, conformance to process, and estimation accuracy?
- Can you track defects against Quality targets? Have you identified Quality targets and are you negotiating and redefining them if project conditions change?
- Are you reviewing Quality targets and metrics at every project meeting?
- Do you understand user needs, wants, and expectations?
- Do you have effective configuration management processes in place and does the project team understand them?

- Does your change control process include: identification, reporting, analysis, and implementation tasks (version descriptions, unique identifiers, and so on?
- Have you documented your plan (including software lifecycle, roles/responsibilities/configuration management plan/quality assurance plan/design and development plan)?
- Have you considered reuse in your development plan and the process by which you will design new components for reuse and reuse existing components?
- Can you track rework costs (and savings when using formal review methods)?
- Have you planned for both white-box testing (development side) and black-box testing?
- Have you built both informal overviews and formal reviews of project documentation into the schedule?

People Management

- Have you trained the staff and provided growth opportunities for them?
- Have you included overtime assumptions in the schedule documentation?
- Have you determined a way for management to be rewarded for increased effectiveness and morale of staff?
- Does the staff understand their project roles and their commitment to the project's success?
- Are you serious about formal technical reviews and is your staff accountable for being prepared for review meetings?

Common Project Management Errors

- Erroneous assumptions; for example, resources unavailable or inappropriate to task
- Missing risk/or risk higher or lower than expected
- Rework costs not tracked/or measured
- Formal review time inadequate

CHAPTER 3

Drawing Your Map— Initiating Your CPI Program

If you can't describe what you are doing as a process, then you don't know what you are doing.
—W. Edwards Deming (1900–1993)

3.1 COMPANY CONFERENCES

Start off your CPI conversion with a company conference at which you brainstorm all the pain points in your organization, group them, and categorize the problems you might be having so that you can identify at which you need to make the first improvements.

We conducted one such conference at the onset of our improvement effort. After we had a few "meet the customer" sessions during which we brought in a cross section of customers for discussion on what they liked about our products, we organized a miniconference. We used our first set of champions to help facilitate this session, and each person had a station in their area that they manned and facilitated.

We started with our mission statement, which was aligned with corporate goals. From our initial meetings with customers and our observations about the organization, we determined a few areas that were likely candidates for improvement. We then distributed a simple survey so that we could get a good baseline on what the workers were experiencing.

In order to plan your own miniconference, you will need to determine a rough schedule and plan for gathering information. It is very enlightening to talk to customers to get their impressions of and wishes for your products. Distributing a survey is not as hard as it seems. You first must determine what question you are trying to answer. In most cases, the question to ask is, "How can we improve our business so that we understand our customers so well that we can deliver a product that they will flock to buy?" Once you understand what questions you are trying to answer, you will want to design your survey to get to the heart of the issues at hand. You need to be careful to keep your survey to 15 questions or less in order to keep the focus on what you are trying to answer. This will also encourage respondents, because the survey will not take very long to complete. Participation is important, and you should drive people to respond to your survey in whatever way you can. Have the survey distributed by the Chief Executive Officer and cosigned by all the Vice Presidents. In this way, the organization will see that this is important, and people will be likely to respond. If you do not already have a mechanism for survey distribution and collection, you can use a relatively inexpensive and simple one from such as *SurveyMonkey.*

The things you need to do to prepare for a miniconference are:

- Gather information from your customer
- Determine what your mission statement for this effort is
- Align your future CPI effort with an organizational goal
- Summarize potential areas for improvement and gather input from your workers on those areas

Once we had our baseline information, we identified initial areas of possible improvement. We then launched a miniconference that took only 90 minutes to complete. The software quality engineering personnel were the conference facilitators who took the lead on both presenting the overall information and moving the session along on schedule. It is very important to have at least one facilitator at this session; the lead facilitator should have some experience in working with large groups. We started our miniconference with a discussion on the survey results and explanations of what we felt were good starting points. We then set up several interactive stations (with giant tear sheets) with particular areas of interest identified (e.g., system design, system requirements, quality, configuration management, project management, and a miscellaneous topic). After that, the conference attendees were tasked with putting subjects in each topic area, and were given a 30 minute time period to complete this. We included a miscellaneous category so that we could collect items that did not fit anywhere else. We then summarized the results for the entire group and had a great starting point for CPI efforts.

This method can work with small groups to large groups; for very large groups, coordination of several sessions might be necessary. Also, in a large organization you will need to be cognizant of any political issues, and engage with additional people in the planning process to ensure that all areas of the organization understand what the purpose of the session is as well as the expected outcomes.

Items you will need for a miniconference are:

- A very large conference area.
- Poster boards—large boards on easels for writing. Assign one subject area per easel.
- Markers—for clearly identifying each subject and for helping to group ideas and mark important points.
- "Sticky notes"—several pads for workers to make their notes. You can get different colors for each topic if you want to.
- Printouts of each topic area,with information starters.

At the end of the brainstorming session, the area leaders summarize the main points in each topic area to the group. After the conference, the software quality engineering team should summarize and distribute an overall summary to executive management. This, then, forms the basis for the starting point of the first big CPI win, and the first set of teams are ready to begin the effort of determining how to move forward.

3.2 DISSEMINATING THE INFORMATION

In order to communicate goals effectively, you must get the correct information disseminated as quickly as possible, and to the right people. The information often moves slowly or not at all between the improvement groups and the people in the trenches. The grapevine might not reach across your forest, and those engaged in one effort may not know that another effort even exists. Some groups see things as changing, but others say, "These improvement efforts haven't reached *me*."

This problem is the driving force behind establishing software configuration management, software development, and software quality assurance plans. Both the Software Engineering Institute and the International Standards Organization models require the establishment and maintenance of such plans. These documents contain the schedule and provide a means of activity coordination. Each of these plans should discuss the authority of software quality engineers and advocates. To ensure that all members of the team understand this, the Program Manager, Project Manager, and Functional Managers should be required to review and approve each of these plans.

Other methods to remedy communication shortfalls include:

- Weekly forums
- Project team meetings
- Reporting

3.2.1   Weekly Forums

Meeting with all members of the Advocate Program on a regular basis is essential. Initially, this should probably be on a weekly basis. It takes several meetings to brief and train the group on their roles and responsibilities, document process areas, and perform assessments, surveys, and reviews. Holding the meeting at a set day and hour each week helps solidify the foundation of the program. As individuals become more involved with the Advocate Program and recognize its effect on their own program, the level of questions and requests for future forum topics will surprise you.

At your company, you can test the tolerance of individuals by conducting the weekly meetings during lunch. Add a little extra in-

No boxed lunches for us; we served home-cooked meals! Twenty to thirty champions and advocates and their appetites attended each weekly working meeting. What came as even a bigger surprise after only the second meeting was the willingness of the champions and advocates to prepare lunches. It was also a great way to share cultural and religious difference within the team. Over time, the cultural culinary diversity evolved into a contest to determine the best way to prepare beef brisket or pork roast. Many team members confessed that they would not have come if it were not for the lunches.

centive each week by providing homemade hot lunches. Surprisingly, they may enjoy this thoroughly!

The investment in making these weekly forums enjoyable and fun will pay off in process improvement results after only one year.

These forums are also an excellent opportunity to model how you should conduct meetings, in general, and process meetings, in particular. Just like you need to prepare a child for what will happen at the doctor's office, so will you need to prepare your group for what are the expected outcomes of each meeting. We say doctor's office because, just as you go to the doctor to prevent or diagnose problems, you will be using these forums to prevent and diagnose current problems in your processes.

3.2.1.1 Chartering

In the first meeting, you should review the project charter. The charter should include all the information discussed in the following paragraphs.

3.2.1.1.1 Project Name and Charter Dates. The project name should be something succinct that identifies the overall project. It could be something as simple as "Company Process Improvement Project," or something more interesting such as "The Software Process Improvement Mission." Try to make it sound like fun and important so that the name alone generates interest and buzz. It is also important to identify start and end dates. We know we said process improvement efforts should be continuous and never end, but a project charter should have defined target dates. If you leave it open-ended, then the project will get "fuzzy," similar to having fuzzy software requirements, and you will lose focus on your project and not deliver what you intended. It would be like having an open date to deliver a product; you could go on forever perfecting the product or adding new features, but you will never actually deliver. You have to be able to achieve a goal in order to declare success. With process improvements, you may have several phases. Phase 1 would be creation of templates and finalization of process definitions. Phase 2 could be improvement and targets for specific improvement goals.

3.2.1.1.2 Project Sponsor. Your project sponsor should be someone in executive management. You will need to identify the person who can have an effect on corporate performance goals.

3.2.1.1.3 Business Sponsor. This person should be someone who can attend your process forum meetings regularly. It is important that they be visible and respected by the organization, and have the ear of executive management. This sponsor should be the one who communicates progress to the executive management team.

3.2.1.1.4 Team Members and Their Roles. These members should understand the amount of time that will be required of them as well as their roles in the teams. You should identify their functional areas, and they should have a specific role and title (e.g., Software Configuration Management Champion). By posting this charter, your organization will know who the "go to" people are and their roles. It also clarifies what is expected of them.

3.2.1.1.5 Vision and Mission Statements. Tie this in with any business objectives you might have (e.g., from a yearly business scorecard). The vision statement should be the long-term goal for your project, the 3–5 years vision of what the company will look like when you are successful with your project. The mission statement should be the mission of the project in the time frame you have identified in the date field. Often, you can find an inspirational quote that sums up the project, and you can put that quote above your vision and mission statements as a way to focus your team on what you are going to accomplish. Your mission statement describes what business you are in and who your customers are. As such, it captures the very essence of your enterprise—its relationship with its customers.

Developing your mission statement is the step that moves your strategic planning process from the present to the future. Your mission statement must work not only today, but for the intended life of your strategic plan, of which your mission statement is a part. If you are developing a five-year strategic plan, for example, you would develop a mission statement that you believe will work for the next five years.

Focus is a primary benefit of your mission statement. It should be broad enough to allow for the diversity (e.g., new products, new services, new markets) you require of your business. It will also be specific enough to provide the focus necessary to the success of your business. Below is an example of a mission statement:

> Clayton Instruments Company designs and manufactures
> highly reliable monitoring equipment for use in harsh or
> unusual environments within the process industries.

Note that this mission statement has both an internal and an external dimension. Internally, it describes the products that the company offers: "highly reliable monitoring equipment." And it lists the functions the company performs: "design and manufacturing." The mission statement also includes the necessary external dimension. It identifies the customer: "the process industries." And it cites the company's "market position"—the reason why customers would prefer to buy products and services from the company. Specifically, the company's products are "for use in harsh or unusual environments."

John Bryson, author of *Strategic Planning for Public and Nonprofit Organizations,* states that, typically, a vision is "more important as a guide to implementing strategy than it is to formulating it." This is because the development of strategy is driven by what you are trying to accomplish¯your organization's purposes [16]. A mission statement answers these questions:

- Why does our organization exist?
- What is our business?
- What values will guide us?

This is because a vision is not true in the present, but only in the future.

Your strategy team will need to develop a compelling vision of the future, a vision that your employees enthusiastically embrace because it challenges them to grow. Let us consider an example of a drive-up window service at a fast-food restaurant. The vision might be one of "Saving Time for Busy People." See? A big vision! And it provides direction for the employees. Some years ago, a client of ours remarked, "Our employees are eager to feel a sense of passion; it's up to us to tell them what to feel passionate about." That's what the vision is all about.

In order for you to get your employees to be passionate about your vision, it has to be compelling. It has to matter, not just to your management team, but also to your employees. To "triple sales revenue next year" does not do it. This only affects a few people. A vision statement should make a difference to your customer, to your community, to the world. A great vision statement might improve the lives of human beings. That matters!

Your vision statement should project a compelling story about the future. When Steve Jobs said, "An Apple on every desk," there wasn't an Apple on every desk. In fact, there won't ever be an Apple on every desk. That's OK. The vision can be figurative, rather than literal.

It's also important for management not just to speak the vision, but also to live the vision. Apple Computer did this. Did you know that the entire design team for the Apple II GS computer signed their name on the artwork for the computer's motherboard? On each and every Apple II GS computer, the team's signatures appear in copper script. Now, that's involving employees in living the vision.

Your team needs to decide how it will communicate its vision to the employees. Decide how to nurture and support that vision every day, in every way; and empower employees to fulfill that vision.

In short, a vision statement should challenge and inspire the group to achieve its mission. As part of the process, you may brainstorm with your team regarding what you would like to accomplish in the future. Talk about and write down the values that you share in pursuing that vision. Different ideas do not have to be a problem. People can spur each other on to more daring and valuable dreams and visions, dreams for which they are willing to work hard. The vision may evolve throughout a strategic planning process, or it may form in one person's head in the shower one morning! The important point is that members of an organization without a vision may toil, but they cannot possibly be creative in finding new and better ways to get closer to a vision without having that vision formally in place. Organizations with many of their staff actively looking for ways to achieve a vision have a powerful competitive and strategic advantage over organizations that operate without a vision.

3.2.1.1.6 Project Strategies and Goals. Strategies are the broadly defined four or five key approaches the organization will use to accomplish its mission and drive toward the vision. Goals usually flow from each strategy. Some examples of a few strategies are employee empowerment and teams, pursuing a new worldwide market in Asia, and streamlining your current distribution system by using lean management principles.

Below are more examples of common strategies:

- **Expand** the customer base and enhance the franchise by pursuing multimedia opportunities.
- **Deliver** an award-winning level of journalistic excellence, building public interest, trust, and pride.
- **Provide** vigorous community leadership and support.
- **Instill** an environment of internal and external excellence in customer service.
- **Empower** and recognize each employee's unique contribution.

- **Achieve** the highest standards of quality.
- **Improve** financial strength and profitability.

After you have developed the key strategies, turn your attention to developing several goals that will enable you to accomplish each of your strategies. Goals should be SMART: **s**pecific, **m**easurable, **a**chievable, **r**ealistic, and **t**ime-based.

3.2.1.1.7 Team Chair and Cochair. There should be a stated chair and cochair of the team. This is generally the highest-ranking software quality or business process person you have in the organization (e.g., the Director of Quality or the Director of Business Process). A cochair should be either a dedicated software quality engineering employee or a champion. It is important that the organization respect the identified people, which enables the facilitation of effective meetings. The team chair will be the tiebreaker in any disputes over priorities or process issues.

3.2.1.1.8 Business Objectives. Document what specific business objectives this effort will affect and how it will affect them. Since process improvement efforts often are hard to measure in hard dollar returns, it is imperative that you figure out how the success of your project will affect the bottom line. This will help you determine both your success criteria and some possible metrics that you can track. Compliance metrics are important, but more important is the ability to show a return on investment for your particular improvement project.

3.2.1.1.9 Resources Needed. Along with the business objectives, it is important to show what the investment in the project is likely to be. To be sure, there will be person hours associated with working on the improvements. There is time involved in just the meetings. Even if you are successful in implementing lunch meetings, there will still be work to be done outside of those meetings. When recruiting team members, you will need to let them know exactly what time commitments they are signing up for. Documenting this will make the time investment more concrete on both sides. The team member will know and understand how much time he or she will have to set aside for the project, and you will know if you start to exceed the expected time investment, while adjusting the project accordingly.

There may also be other hard costs, such as investment in tools and/or special equipment. For example, you might need a good defect-tracking tool and you are currently using spreadsheets. You need to evaluate where you are as an organization and make a best guess on costs to make improvements. You do not have to express these as specific dollar amounts; you can merely list additional tools/equipment and cost to be determined (TBD). At least you will have documented that there may be some tool costs.

3.2.1.1.10 Limitations and Dependencies. This section describes dependencies in the organization, and clearly states what the team does and does not have control over. A constraint may be that no more than $50,000 is available for the effort. A boundary example is the breadth of authority of the team or how the team will interact with executive management. A boundary statement could be "Team will recommend specific performance objectives for annual reviews." Further explanation of this boundary would be that the executive team will review the objectives and make the final decision as to how objectives will be worded in annual reviews.

3.2.1.1.11 Stakeholders. These people have an investment in the project. Typical examples are customers, shareholders, and engineering, management, marketing, and customer service personnel.

3.2.1.1.12 Measures and Metrics. These measures will tell you that your efforts are effective and that you are helping the bottom line. It is very important to identify these early on in the project so that you can baseline where you are before you begin to make improvements. If you wait until you are six months into the project, you will have to produce appropriate historical data so that you can compare before and after results. Sometimes, measures can temporarily go up. For example, placing an emphasis on the design phase of product development may increase the amount of time the team spends in development. Be aware that you may see an uptick in investment before you see an increase in returns. Try to find at least one metric that you can claim a win on right away. People want to be on winning teams, so be sure you find a way to show a win early on in your process-improvement activities.

3.2.1.1.13 Importance of Agendas. You will need to send out an agenda at least 24 hours prior to each meeting. Since your meetings will be recurring each week, select the topics to be addressed ahead of time and discuss the upcoming topic prior to adjourning. By doing this, everyone will be prepared and can keep a special focus on the upcoming topics. This also provides an excellent opportunity for the members of the organization to identify their current pain points. By listening to what is important to them, you can prioritize how you will relieve the most important pain points. This also engages the team in troubleshooting and finding issues while they are still small, before they have a huge impact on the end product. By engaging with those actually doing the work and responding to their needs, you will inevitably be validating that you care about what the workers think.

The meeting agenda should be time bound per discussion topic with an identified leader and objective. Your organization probably already uses a meeting template, and you can adapt that to your process forums. You may even discover a needed agenda improvement as part of your forum activities.

3.2.1.1.14 The First Meeting. The first meeting will be to review the charter previously created in the weekly forums. Before you can have that meeting, though, you need to review with your sponsor a draft vision and mission statement that you propose for the charter. You should also discuss the charter at a very high level with both the team chair and a few trusted advisors. Your sponsor should then discuss the proposed charter with executive management to get buy-in at the proper level to help drive the effort. For this to work well in your organization, you will have to approach the effort from both ends. Executive endorsement of the effort will get things going and get some accountability in the organization. Without that endorsement, the workers may or may not engage in the effort, depending on how passionately they believe in the effort, or on how much pain they are experiencing because of the current process.

You can then conduct your first meeting and introduce a draft charter. You should present this as a draft only to get the team started on brainstorming on what the details of the charter should be. In working with teams, it is very interesting to watch the dynamics of how the charter evolves in the very first meeting. At first, the team acts as if it is a silly exercise to agree on the vision and mission statements, and then someone makes a cursory "it looks okay to me" statement, and then there is a lot of head nodding. Then, inevitably, one person will ask, "what do you mean by project artifacts? Is that just code?" And the discussion takes off.

It is not unusual to spend an entire meeting on just the vision and mission statements. Do not be alarmed if this happens. Once you have those two statements and maybe a quote that helps the team focus, you are well on your way. If you have a good meeting facilitator, then you will get through the entire charter in one or two meetings. If you are taking longer than this, then you are probably spending too much time word crafting.

Once you get the finalized project charter finished, and once the team agrees on it, you should go back to your team's sponsor to present the chapter to the executive team for awareness and final approval. The level of detail that you present to your executive team will vary depending on your organization's communication style. Your project sponsor should be able to advise you as to what level of detail you will present.

You will want to keep the information flowing to the executive team at a high enough level to allow room for the newly formed core team to craft the majority of the charter. This is part of the ownership of the project. If you just hand down an edict from above, your team members will have a difficult time owning the overall project. By engaging the project team by first writing the details of your charter, and then securing executive sign-off on that charter, you have the bookends you need to ensure success of the project. The rest will be up to the team to fill in the missing pieces.

3.2.2 Project Team Meetings

One of the advocates' largest responsibilities is to discuss their activities at their project's team meetings. These meetings are the conduits for sending information from the Advocate Program back home to the project. At your company, during each weekly project meeting, advocates will discuss upcoming assessments and reviews, and will highlight any accomplishments. Having an advocate briefing at the team meetings is also the most effective forum for briefing the project about its noncompliance with their processes. It is essential that you note the advocate briefing on the meeting agenda. The advocate briefing should be an integral part of the project status briefing.

3.2.3 Reporting

A project-independent advocate is not essential initially. You must continuously reassure the advocate that his or her input is always valued. You should communicate both good and bad news to the project at weekly project meetings.

At each meeting, present graphs of the success metrics that you identified in the team's project charter. Highlight your team's contribution to those metrics. This data can be used in report roll ups at the organization level and reported upward. The team can visually see progress, and can comment on what they can do to improve even further. It is also important to show the team how they are doing in the event that they either did not agree or encountered unusual circumstances that might have positively or negatively affected their performance numbers.

If there is an issue that cannot be resolved with project personnel, the advocate may discuss the item with the Quality Manager. It is important that the advocate be empowered by executive management to stop a noncompliant project. For this reason, it is better to share advocates. In other words, you would assign an advocate to report on a project to which he or she does not directly report. In the case of two or more projects, the projects would trade an advocate's time on an independent project. In this way, the advocate can be spared the fear of a backlash from his or her own manager for reporting process noncompliance.

3.2.3.1   Newsletters

In your organization, publish a monthly process newsletter to ensure that the organization knows what they should understand about the process-improvement efforts. The contents of the newsletter will contain the latest metrics and news of any recently made process changes. It is extremely efficient to include a "Tools and Tips" section. Engaging the champions to write content for the newsletter will also help to raise awareness of their key areas, and will advertise who is in charge of each area.

Here is an outline of a typical newsletter:

- Process Headline—any announcements that are new or exciting
- What's New
- Key Process Area Reports
- Tools and Tips
- Note from the Business and Quality Director
- Executive Comments (include a signature, if possible)

This newsletter should be drafted and then presented to the executive sponsor for approval before it is distributed to the rest of the organization. By assuring that the executive sponsor approves of the newsletter, you will receive a few key side benefits: the executive will be updated on the latest news and will agree that the activities were in alignment with business goals. It also ensures that the executive sees the incremental return on investment. You can also share this newsletter with other organizations that might be involved in their own improvement efforts. This sharing could leverage your work, and the other organizations could offer advice and suggestions to you.

3.2.3.2 E-mail

Keep e-mails to a minimum in your organization. At most, keep any e-mail communications to a very short update. Furnish a link to your internal process website with a notice of either changes or successes. When teams are evaluated by the Quality Director, send a congratulatory e-mail out to the organization. You will probably need your business sponsor to send out organization-wide e-mails.

CHAPTER 3 SUMMARY

Key points from this chapter on gathering and disseminating information follow.

Weekly Forums. Gather all champions and advocates regularly, and make the forums open and fun. Use these as opportunities to model how they should effectively conduct meetings.

Chartering. A charter will help keep the teams focused and aligned with business visions and missions. Elements of a charter could include:

- *Project sponsor*—preferably a member of senior management.
- *Business sponsor*—your main communicator with executive management.
- *Team members* and their roles, particularly if they are assigned a key role in a particular functional area.
- *Vision and mission statements* that can be used as guiding principles and help in decision making and direction setting.
- *Project strategies and goals*—four or five key approaches that can support achieving a mission and drive toward a vision.

A charter will also identify who the team chair and cochair are as well as business objectives, any resources that you may need, limitations and dependencies, stakeholders, and success metrics.

Agendas should clearly state expected outcomes of each segment of a meeting with outlined times and leaders for each topic. Capture action items, and identify the owners and target-resolution dates.

Project Team Meetings. These meetings give the advocate an opportunity to give ongoing feedback on the health of the project in terms of the objectives upon which the CPI teams have agreed. Project team status can be discussed in just a few minutes at each meeting.

Reporting. Reports can include graphics of each success metric that you identified in the team charter. Roll these reports together as an aggregate to show the overall group status.

Newsletters. These are either electronic or printed materials. Use them to communicate status, news, and successes. Ideally, the Chief Executive endorses this newsletter.

E-mails. Use these for short updates and congratulatory messages. Always include a hot link to your internal process web page if you have one.

CHAPTER 3 AIDS—AN EXERCISE IN FORMING VISION

This section outlines an exercise you may employ to assist your organization in defining its own vision. By using this exercise to develop your organizational vision, you may be better assured that the vision statement that is developed is a shared vision.

At a retreat, or even at a board meeting or staff meeting, take an hour to explore your vision. Breaking into small groups helps increase participation and generate creativity. Agree on a rough time frame, say five to ten years. Ask people to think about the following questions: How do you want your company to be different? What role do you want your organization to play in your company? What will success look like?

Then ask each group to come up with a metaphor for your organization, and to draw a picture of success, such as, "Our organization is like a mariachi band, all playing the same music together, or like a train, pulling important cargo and laying the track as we go, or. . . ." The value of metaphors is that people get to stretch their minds and experiment with different ways of thinking about what success means to them.

Finally, have all the groups share their pictures of success with each other. One person should facilitate the discussion and help the group discuss what they mean and what they hope for. Look for areas of agreement, as well as different ideas that emerge. The goal is to find language and imagery that your organization's members can relate to as their vision for success.

Caution: Do not try to write an initial vision statement with a group. Groups are great for many things, but writing is not one of them! Ask one or two people to try drafting a vision statement based on the group's discussion, bring it back to the group, and revise it until you have something that your members can agree on and that your leaders share with enthusiasm.

CHAPTER 3 AIDS—SAMPLE WEEKLY FORUM TOPICS

 Generally, we use two very specific agenda types: the time-bound agenda and the topic-focused agenda. The topic-focused agenda is more open and leaves more room for adjusting timings based on the discussions. This is typical of a process-improvement agenda. Team-meeting agendas are a bit different for us in that an expected outcome is normally stated. By using the agenda below and keeping the topics more free-flowing, we built in a sense of openness in the resultant discussion. The agenda sample below is a listing of actual topics from a typical weekly forum meeting.

Agenda Topics

Implementing SEI model: KPA champions will present action items for moving to the next level.

	Name	Time

Introduction of agenda and latest updates:
    Current status
    Compass polling
	Celeste	

Finalize CCB process discussion	Jeff	
Software engineering model training on compact disc—an overview and instructions	Celeste	

Wrap-up:
    Decision on next week's work actions
    Upcoming events; cross- assessments
	All	

CHAPTER 3 AIDS—CHARTER TEMPLATE

Use a charter template to kick off a new project and help organize a team around a central purpose. As it is much less formal than a cost–benefit analysis or a Six Sigma project, you can use a charter to jump start a team's efforts. Note that there is a specified end date. This is not to imply that improvement efforts ever end. It is good to have a defined date to check back in with the project to see if it needs to be refreshed with a newer charter. In this way, efforts will be living and growing. A sample charter template follows.

Team Name:	Start Date:	End Date:
Team Chair:	Team Cochair:	Sponsors:
Mission Statement:		
Project Objectives	Team Members	Area represented/role
• Objective 1	•	•
Linkage to Corporate Objectives:		
Limitations/Dependencies:		
Resources Required:	Cost:	
Stakeholders	Organizational Impact Expected:	
Metrics	Description	Owner

CHAPTER 3 AIDS—METRICS

The following list of metrics could be used to track project compliance and success. Choose those metrics that make the most sense for your organization. Try to not make your process tracking too complicated. If you are not using a metric, leave it out.

- **Compliance Ratio.** Number of total projects in organization divided by number of projects in compliance (measured at 80% compliant).

- **Average Compliance Level**
  - By project: Percent of compliance per key process area on each project. Average all scores (percent compliance in each key process area divided by total number of process areas tracked).
  - By organization: Average of all project averages.
  - By key process area: Average all the scores in each process area by the number of projects that have been scored.
- **Requirements Churn.** Number of requirements changes from original, approved requirements document(s).
- **Defect Density.** Number of defects, minus number of errors, divided by all defects. (Errors are defined as faults found in the phase in which they were created; defects are faults found in phases after they were created.)
- **On-time Delivery.** Number of projects delivering on time (as planned).

CHAPTER 4

World Vision—Training the Organization

Human nature is not a machine to be built after a model, and set to do exactly the work prescribed for it, but a tree, which requires growing and developing itself on all sides, according to the tendency of inward forces which make it a living thing.
—John Stuart Mill, *On Liberty*

4.1 ENGAGING PERSONNEL

Training non-Quality personnel could be easier than it sounds if you consider that the training is part of the big picture, or the overall vision for your company. First, you need to realize that you have to instruct members of the organization in the defined process anyway. If you are not doing this already, you should start now. Everyone in an organization needs to understand how things work. You will be adding instructions into your everyday processes that will encourage a process mentality.

It is important to leverage the required training by engaging the organization early in process definition and by using pilots, including quality tools. This allows the staff to begin to understand the processes they need in order to be able to perform process assessments in the future. In fact, the personnel have bought into the new processes through the pilots, because they were actively engaged in them. They had a say in how they might modify the processes. In allowing them to take ownership on several levels, you will build in the idea that everyone is responsible for Quality.

4.2 MATCHING THE TOOL TO THE LEVEL OF QUALITY MATURITY

Quality tools are the mainstay of Quality process development and analysis, and you can gain competence in the use of quality tools through many avenues, including reading, seminars, and hands-on experience. What may be more difficult to learn is whether a particular tool is a good fit for the degree of quality maturity of the organization. As part of your team-building exercises, you may find it helpful to review several maturity models with your teams. Through this, you can fit the right maturity evaluation and tool to your particular organization and unique situation.

Different authors have proposed several maturity models over the years. Philip Crosby proposed his five stages—maturity uncertainty—in *Quality is Free* [17]. The stages consider major issues, such as level of management understanding, handling problems, and cost of Quality performance results.

In *Levels of Maturity Matrix,* Russell Westcott proposed a similar model and named his stages dysfunctional system, awakening system, developing system, maturing system, and world-class system [18]. As with most such models, its intent is to allow organizations to determine their status and monitor their progression.

In the book titled *Critical SHIFT: the Future of Quality in Organizational Performance,* Lori Silverman's hierarchy of five fields of performance practice consists of quality assurance, problem resolution, alignment and integration, customer obsession, and spiritual awakening [19]. The model begins with a narrow and reactive view of quality and evolves into the idea that quality of society is the ultimate goal. Our collaborative model fits nicely in with the idea that quality of society is the ultimate goal of performance. It is interesting to note that we are also addressing the ability of your organization to engage in processes that will reflect an obsession with its end user, the customer. Your job will be to determine what sort of guideline you will use as your own.

Maturity levels have also shown up in quality awards programs. The Tennessee Quality Award model is based on four levels: quality interest, quality commitment, quality achievement, and quality excellence. It enables organizations to receive recognition commensurate to their degree of effort and success toward total quality.

The tools an organization uses will depend partially on the organization's understanding of Quality. One could also say that the tools used, and those ignored, may limit the degree to which an organization can achieve a higher level of quality performance.

Table 4.1 summarizes the levels of quality maturity and the tools that might be more appropriate at each level. Boundaries be-

Table 4.1. Maturity levels and Quality tools

Maturity	Description	Useful Tools
Low	There is no formal Quality system, or if one exists, you do not practice it on a daily basis. Customer complaints and other external failure costs are high. The Quality Department is held responsible for quality. There is little or no formal continuous improvement involving people who actually carry out work processes.	Seven basic tools Audits Cost of Quality Statistical process control
Medium	External failure costs have been lowered, but there is till a high rate of internal failures. Each department accepts its role in the Quality management system, and improvement projects involving employees are frequent.	Creativity tools Customer surveys Failure mode and effects analysis Benchmarking Design of experiments
High	Organizational strategy seamlessly integrates and drives management systems for areas such as Quality, safety, environment, and finance. All departments and work processes monitor their own performance and implement daily improvement aligned with strategic direction.	Seven management tools Employee surveys Quality function deployment

tween maturity levels are not completely clear, and a wide range of exception may exist based on industry, management expertise, and the previous history of success of the organization.

Whatever future industry standards you might be asked to follow, including models developed at Carnegie Mellon University's Software Engineering Institute and International Organization of Standardization, or any of a number of other guidelines, including those provided by Institute of Electrical and Electronics Engineers, Inc., you will want to evaluate the standards for application to your specific needs. The important thing is to start with some sort of framework, and use the quality tools that will assist you in delivering processes in that framework. If you choose a well-known standard, it can help motivate employees because they will be able to add that to their resume. Although this may not seem like a wise thing to do, it is motivating to an employee to be able to reference known standards on their employment history. The reality is that few jobs last forever in the current economy, and resume building is an ongoing hedge against unemployment.

Although you may have several individuals in your company who profess knowledge of business processes and Quality, you need one (or a few) person who really understands the complexities of business processes and Quality. Typically, the Director of Quality and Business Processes reports to a Vice President. This person is essential to your success because he or she is likely to have extensive experience in rolling out organizational changes, in addition to having a depth of Quality knowledge that your organization needs in applying any standard framework. '

In order to begin building training modules for advocates, consider building a training program within your organization instead of contracting out.

For advocate training, our first proposal was to subcontract all training to an outside source. We investigated hiring a subcontractor from two local training houses to assist with training classes. Each training house typically charged $36,000 for six classes. The six classes were to be conducted twice in-person in the United States and three times via net-meeting at our remote sites, for a total cost of $180,000. This fee would literally double if the training houses were to tailor each group's specific need. To train all our advocates, we required six classes at two domestic locations and at three remote locations: Europe, India, and Asia.

Fortunately, we discovered that we had the talent in-house to establish our own curriculum, and delivered the six classes noted in Table 4.2 at our five sites ourselves.

The in-house talent can come from the advocates and champions. They are able to translate the standard guideline into terms that the average worker can understand. They also can break down the guidelines into process areas for training. In the case of the Software Engineering Institute's Key Process Areas for Software, they have easily identified the process area descriptions.

It is important to break down training modules into logical, understandable parts, so that it is easy for an employee to learn the

Table 4.2. Training classes

Class Title	Audience	Class Length
Advocate Roles and Responsibility	Advocates	2 hours
Performing Assessments	Advocates	2 hours
Software Configuration Management Plan	Advocates	2 hours
Software Quality Assurance Plan	Advocates	2 hours
Key Process Area Overview	All	4 hours
Key Process Areas	Advocates and champions	8 hours

principles in a time period that is not too lengthy. The training curriculum for each class should be coordinated with and receive approval from your company's training department. Training modules are generally planned by your training coordinator to be no longer than 30 minutes. In this way, the employees can choose the length of time they want to devote to training and can optimize their learning according to their individual preferences.

When developing your training materials, you should route all class materials and attendance information back to your Learning and Development group for archival purposes. This results in curriculum being validated and monitored by the Learning and Development group and judged as accredited training for individuals. If your organization does not have a Learning and Development department, you could have your Quality expert track this information. If this is not feasible, you could also engage your Human Resource department to help and make this part of an employment record.

4.3   USING CHAMPIONS

Once you have designated champions for each of the process focus areas, you can make these key people trainers. Spend time with the champions to ensure that they understand their particular process area. Develop customized PowerPoint curricula for each site in your organization and have your local champions provide voice-overs. When you develop the training modules for your organization, provide the narrative that your champions will use as a guide for their training sessions. You can then record these classes on compact discs. In this way, you feature your champions and get their names and voices out there.

This holds a twofold advantage: you deliver your primary goal of training, but you are also able to publicize who your champions are so that people who have questions or, more importantly,

We also investigated hiring a local firm to record videos that could be used to train future Advocates on thirteen key process areas. This would include customization for each of our global sites: the United States, Europe, Asia, and India. The estimate was $4,000 per training module or KPA, or $52,000 for each of our sites, for a total cost of $208,000. By utilizing our champions, we avoided this cost.

have improvement suggestions, know who the primary contacts are.

By recording your own training modules, you can customize the training to include company-specific support tools and processes, including any site-specific customizations that you need. For example, since your organization may be spread across the globe, some locations may use a different configuration management tool. When you reference which tool to use in configuration-management, substitute the right tool for the right site. This makes the training more effective and engages regional representatives in the process. Oftentimes, your remote team members feel isolated from the "mother ship" and this customization is a way to demonstrate to them that you remember them as key members of your team.

Side effects of Champion Training recordings are:

- Champions become leaders in the group
- Champions receive celebrity status
- Champions experience upward mobility
- Champions are recruited by other projects as specialists
- Being a Champion is required prior to promotions

The result will be a robust, customized training program digitally captured for future use. You can ensure that new personnel are trained in the same way, with the same terminology, and that they will understand immediately who the contact people are and who are considered area experts. Not only will this result in a cost savings in training dollars, but it will avoid the inevitable costs of miscommunication.

4.4 TRAINING WITH APPLICATION

Training sessions in which you follow the conceptual training with the application of the training are the most effective training sessions in which your personnel will ever be engaged. If possible, take advantage of being able to take over a whole wing of a building to accommodate this type of training in a single day. You may

A striking example of this came when we trained the entire organization (sans administration) on our process for formal technical reviews. We had an all-hands event for which attendance was mandatory and we trained everyone in the same review process. We engaged our champions in the training by getting them to perform a formal review skit. We made it fun! The concept training part occurred in the morning, but in the afternoon the real work happened, because we took current software products and broke into teams to practice what we had just learned.

The marketing team was most reluctant to participate in this event, because they did not see what was in it for them; however, by the end of the day, after marketers spent some time with engineers and reviewed a marketing requirements document, they understood. One marketer commented: "I've got it. I will never look at a marketing requirements document the same way again, and I will definitely be writing them differently in the future."

have to take subsets of your groups, depending on the size of your organization.

By including everyday people from various functions in your organization as advocates, you can engage all functional areas in the ultimate goal. They will see the vision of the overall process in terms that are concrete and viable to them, and they will help bring the organization to a higher level.

CHAPTER 4 SUMMARY

We have reviewed how to create a training program specifically tailored to collaborative process improvement. This includes how to choose quality tools and how to match an organization to the tools that supplement given processes.

Outline of How to Build a Training Strategy

- Engage personnel and evaluate tools
- Break down overall processes into key areas
- Determine the key points in each key area
- Build training modules in easy-to-understand language
- Customize (or localize) modules to make them specific to your organization
- Review the training materials with your team members and incorporate feedback
- Provide an easy way to gather feedback on training modules for continous improvement
- Use champions as trainers

CHAPTER 4 AIDS—EXAMPLE OF KEY AREAS, KEY POINTS, AND TRAINING

Key Area: Design

Key points:
- Sufficient reviews
- Standard taxonomy

Training:
- Review training
- Review checklists
- Design standards

Key Area: Configuration Management

Key points:
- Configuration management tool
- CM guidelines
- Build statistics
- Change requests
- Release readiness

Training:
- Configuration management standards
- Configuration items
- Change request format and notifications
- Components for release readiness

Key Area: Software Planning and Oversight

Key points:
- Estimates
- CPI tasks
- Metrics
- Processes

Training:
- Guidelines for estimations
- Policies and procedures
- Project management tool and entering task information

Key Area: Software Quality

Key points:
- Defect density
- Testing and reporting

Training:
- Capturing errors and defects
- Building test cases from use cases
- Standardizing review processes

Key Area: Software Requirements

Key points:
- Requirements churn

- Focusing on the customer
- Management

Training:

- Meet the customer—profiles of product in use
- Requirements definition
- Requirements documentation and traceability

CHAPTER 4 AIDS—TRAINING CURRICULUM OUTLINE

Here is a brief outline of how to put together your training program:

- Break your overall process down into easily understandable process areas, by phases if you have phased-based development product life cycles. Many standards, such as the Software Engineering Institute's Models for Software, break this down for you.
- Consider what the important aspects are for each process area. Think about it this way: What would you like your personnel to learn in this section? Ask yourself if this module will fulfill that purpose. consult your business and quality process expert for help.
- Build each module using simple terms that everyone can understand. You may consider splitting training into expert and user modes so that you can go into more depth with some people (e.g., champions) and less depth with others.
- Customize each module with things that are pertinent to your organization. If you are using a configuration-management tool, mention it. Give people additional reference materials within the training. For example, if you have set up a special process repository, ensure that everyone knows where it is by telling them explicitly in the training module.
- Review the module materials in a process-improvement forum or in some other setting, and review it extensively with the champion who is going to do the training. The champions should be engaged very early in the process, because the earlier they begin to help form guidelines, the sooner they will bear responsibility and feel as though they truly own their process area. Be sure to incorporate any feedback you may get back into the training as soon as possible. Assure your team that their input is valued.
- Provide a way to give feedback and suggestions for improve-

ment when you roll out the training materials. There are several packaged training applications that you can look into to ease the burden of building the modules. If you do not need an interactive screen, using PowerPoint with narration is a viable option. Keep it as simple as possible. Perhaps your organization has a collaborative environment for discussion where you can start a blog or threaded discussion on your process areas [20].

CHAPTER 4 AIDS—REVIEW EXERCISE

From a software perspective, we believed that the process of formal reviews, correctly performed, would net us a fast big win that we could use to build a winning case for ongoing change. We decided that in order to emphasize the importance of change, we should have an all-hands meeting and training session that emphasized the need for global change, and that the first step would be to institute consistent formal reviews of all program deliverables, from marketing requirements documents to coding.

One of the things we did to keep it fun and, at the same time, practical, was to set up two skits of teams performing reviews. We used some of the comments we had actually heard in team meetings to illustrate our points. We picked a team of six well-respected individuals to be our actors, and we loosely scripted the session for them. We set them up at a review table in the front of our huge conference center, put microphones on them all, and sat back to watch the fun.

The review started very lazily, with no set agenda, and with one or two team members showing up late. The late arrivals caused lost time in the meeting because everyone stepped back and had to bring the late persons up to speed. The funniest parts of the review were the comments about "I just want to do my work; I don't have time for this nonsense!" We also poked fun at the people who got hung up on a style issue. They would say something like, "I

wish this was in Times New Roman instead of Arial." We also demonstrated why a manager should not attend a review and what happened when he did. We had him articulate what everyone fears in that situation with the words from the manager, "Now, I finally have a way to put your poor performance down on your review!" We ended the "bad review" skit with a question-and-answer period to determine how many of our people could identify with the bad review and point out what really bothered them about the review sessions.

We then broke off and did some specific training around what we considered to be a good review. Each time we made a point that reflected back on our bad review skit, we would point out the reasons why the recommended guidelines were the way they were and could easily illustrate what happened from a team perspective when someone did not follow the guidelines. In this way, we could get the workers to see and feel why they needed change. The exercise was funny, but that was largely because we were all watching ourselves make mistakes and get locked into places where we were helpless as an organization to change things.

After we completed the training session, we modeled a good review by performing a similar skit, only in a good way. The interesting thing about the good skit is that it was not funny! It was crisp and to the point, and the workers got to see a very boring (but effective) team meeting to review a deliverable. It visually demonstrated how things look and feel when a team pulls together and follows a guideline.

By the way, we captured those skits on video and showed them around the world. It was great fun seeing managers and workers interacting in a lighthearted way to demonstrate how destructive a bad review session could be and how constructive a good review session could be for the organization at large. After we completed the session, we moved immediately to the hands-on sessions so that people could

apply what they just learned. We performed real work on real deliverables, and saved the company more than triple the cost of taking them out of the work force to perform this training. This was perhaps the single most effective training that our organization had ever experienced. Even the crustiest of software developers saw the value in the training and began to use it the very next day in their work. We believe they did this because they achieved the improvement in a small period of time and were able to directly apply what they learned to their everyday work.

CHAPTER 5

World Views—Addressing the Capital Q

Quality is a very personal obligation. If you can't talk about quality in the first person . . . then you have not moved to the level of involvement of quality that is absolutely essential. . . . You must be a believer that quality is a very personal responsibility.

—Robert Galvin

5.1 PROCESS OVERVIEW

One of the first things you will notice when working with non-Quality personnel is that they often confuse software assessments with audits. Assessments are very different from audits. Assessments do just that—they assess where the organization is in relationship to its stated guidelines. They can have scores attached to them and are generally a barometer of where an organization is from a process perspective. Audits, on the other hand, are examinations against standards to check for accuracy and compliance. This is a subtle difference, but an important one. No wonder it is confusing!

Because it is confusing and because International Organization of Standardization audits, in particular, can shut an organization down, your organization must learn the purposes of your assessments. You must eliminate the fear factor in order to get on with business. Initially, your advocates may fear team members will see them as the police and the "bad guys" of the project if they take on a Quality role.

For example, "Get out of the way of the policeman that is going to report on us and make us do some bureaucratic nonsense!" is how most team members see processes. For this reason, most non-Quality personnel are very hesitant to take on an assessment role. In fact, in your own organization the Test group may initially carry the overly heavy burden of being the initial advocates. After all, the organization may see them as the police anyway, so why not add the term "process police" to their job description?

5.2   MITIGATE THE POLICE PROBLEM

Approach the police problem with a bottom-up perspective as well as a top-down perspective. Host weekly training and process forums, and keep the atmosphere open and inviting. Depending on the size of your organization, you can either host one forum or split your forums into several sessions. Either way, it is essential to get the organization engaged in a nonintimidating forum for discussion. Ensure that you engage management sponsors in the process and insist that they attend your open forums. In order to recruit future advocates, you must define and train all personnel on the basics of your chosen standard. By gathering champions and advocates together in an open forum, you begin to build a team that understood the common goal of process improvement.

5.3   BUILDING CHECKLISTS

Once you have your process areas defined, the right people recruited, and your communication paths identified, it is time to create your checklists. It is always good to start with an example. For each process area, it is good to start with a straw proposal for that area. It is very interesting to watch the group dynamic when you start with a draft checklist. In general, the first reaction is that it is just fine. Then someone will inevitably ask a simple question like, "Does configuration management also cover documents or is it just for code?" This will have a ripple effect where other questions are asked by the improvement team and then the checklist will begin to be refined based on those questions. This will continue until you have perfected the checklist.

   If you conduct review sessions in the process forums, then you will be able to work through one checklist per session. Remember, you should be able to train each process area in a half-hour period. By keeping the topics crisp, you will be able to work through each process area in an hour's time if you start with a draft and if you have the right people in the room to make decisions. Even if you do not get the checklist right the first time, you will be able to quickly refine it as it is being used. Welcome requests for refinement; however, just as in formal technical reviews, you do not want to wordsmith to perfection—the same goes for checklists. As long as the items on the checklist are understandable, do not allow extended nit-picking over wording. Ensure that you cover the essence of the process area, without spending an inordinate amount of time on style issues.

To create your first draft, review the process areas you have identified and think about what components are important. You will not have to ask a question for each area; just ask questions that would be key indicators of success in that area. You will want to keep the checklists relatively short—no more than 12 questions per checklist. This will force your team to focus and understand what is important to success. Remember also that you are building processes that result in a high-product quality, so your checklists should help you build in quality at every stage.

Here is a summary of the steps in building your checklists:

- Identify process areas that you can train in a 30 minute segment or less.
- Review each process area and select what would be key indicators of success in that process area.
- Draft no more than 12 questions per process area, keeping a focus on product quality.
- Review the draft checklist with the key personnel that you have identified.
- Refine the checklist and route it for final review.
- Repeat the above steps for all process areas.

5.4 PILOTING AND REFINING CHECKLISTS

By using checklists, you will ensure that you are holding all programs to the same process criteria. Even with the guidelines to follow, you will still need to provide a way for others to follow the guidelines while continuously tracking as a part of the daily life of your organization. For this reason, your checklists should be easy to understand and report on in order to help your advocates track their particular areas.

Of course, one method to validate a new checklist is by applying it to a pilot project. Typically, the pilot project is an effort that is well staffed and on schedule. Picking a project such as this ensures that the checklists are well received and executed. However,

We developed checklists in our process-improvement forums sessions. Usually, the Director of Business and Quality Processes developed the draft checklist as a starting point. We extracted the draft checklist from our standard process guideline. We determined what actions, if performed, would be a good indicator that the process was being followed by the group. We projected this first checklist on the wall for all to see and then went to work on it during one of our improvement forums. Generally, the champion would lead the session with his or her area's checklist to further emphasize that the whole organization owned the checklist as part of the tactics we were employing.

For regional representatives, the Quality Director would conduct a web-based session that would mirror results of the local session. Then additional customizations could be agreed to by the local team and the Quality representative. If necessary, a team would make a special checklist for their particular site, depending on how extensive the customizations were.

you can achieve the biggest return on effort if you apply the new checklists to a problem project after you see a process as adding value. Use the buy-in of a problematic program to quickly, easily, and constantly display improvements. Witnessing the improvements on their own efforts can even revitalize project staff.

Transforming a problematic project into a model project is what everyone wants to do. The bottom line is that no one enjoys being part of a losing team. The team will work hard to bring a project back in line, but it must be an attainable goal. You must select the right project, one that is not too complex. The project must be curable; otherwise, you will jeopardize the entire project.

Checklists are the key to smooth, consistent execution. Having a standardized checklist will remove most of the human personality from the assessment equation at your company. While the Quality group initially develops your checklists, the advocates will quickly update them to make them more effective and efficient. During the first six months of the advocate program, they may update each checklist three times on average. That is continuous improvement! Each time the improvement team updates a checklist, it must be reviewed and approved by the entire Advocate team. If your team is not requesting updates, this is a sign that they are not engaging in the process either because they just don't understand it or they do not think there is anything in it for them.

5.5 THE CAREER PATH OF THE ADVOCATE

Convincing management that advocates are management material is one of the first steps you will undertake in order to raise awareness of the critical role the advocates play. After a bit of training, management will buy into the idea and realize that advocate personnel understand the overall process better than most. Once managers see the advocates in action, they understand what value these individuals bring to overall business effectiveness.

CHAPTER 5 SUMMARY

Remember that you should leverage your champions and integrate the vision of your overall process so that your CPI efforts are seamless and constant. Part of how to make the efforts consistent is to set examples for dialogues and also to use repeatable methods guided by checklists.

To Address the Capital Q:

- Mitigate the police mentality
- Engage in friendly and open training sessions about the overall process
- Get visible upper-management sponsorship for the advocate program
- Train everyone on the basics
- Build checklists
- Start with a draft and open a discussion on each checklist
- Review in open forums regularly
- Identify possible checklists by succint process areas
- Identify key indicators of success in each process area
- Keep checklists to a single page or 12 questions or less
- Pilot and refine checklists
- Ensure the career path of the advocate

CHAPTER 5 AIDS—GUIDELINES FOR DESIGN REVIEWS

Characteristics

Prioritize the following list. Know what you expect from the code and where you need to make trade-offs. Prioritizing this list will help define what things are more important and streamline the design process.

High Priority	**Lower Priority**
Security	Elegance
Maintainability	Integrity
Extensibility	Size
Reliability	Portability
Interoperability	Simplicity
Testability	Consistency
Reuse and Usability	Speed

Evaluate the design considering the following characteristics related to quality:

- Interoperability. Will the design meet the requirements for interoperability?
- Extensibility. Is the design forward-looking and generic enough to allow it to support more requirements as time progresses? Consider the road map and future possibilities for the product when evaluating the design.
- Maintainability. Does the design use variables appropriately (e.g., data structures, global, etc.)? Has effective coupling between modules been used or is the coupling too tight?
- Reusability. Check the design for effective reuse of code (e.g., do not tie interfaces to specific scenarios or products if not necessary).
- Efficiency (relates to size, robustness, safety, simplicity, and speed).
- Testability. Ultimately, requirements, design, code, and test should all be linked together to the source of the requirement. A traceability matrix can be useful in tracking the requirement throughout the life cycle.

Technical Considerations

- Data and control flow (with some physical representation). Is the flow of control clear? Are all scenarios covered, including exception conditions?
- Clear and comprehensive documentation. Is the theory of operation clear?
- Is it clear how this fits into the big picture (system view)? Is there a design spec that encompasses everything? Are the interfaces clearly defined?
- Does the design meet all of the explicit requirements? Can you trace the design to a specific requirement and can you trace that requirement to the source, and from the source backward?
- Does the design meet all the implicit requirements desired by the customer? Does it reduce implicit requirements to testable explicit requirements?
- Is output from appropriate design tools available for review?
- Is the design readable and understandable? Can it be used to generate and implement code?
- Does it address the data, functional, and behavioral domains?

Note the most common design errors:

Misunderstanding problem specs
Forgotten cases or steps
Data design mistakes
Inadequate checking
Extreme conditions neglected
Cannot be implemented
Timing problems
Initialization problems
Sequencing mistake
Indexing incorrectly

CHAPTER 5 AIDS—GUIDELINES FOR SCHEDULE REVIEW

Documents Needed at the Review

- Project (hardware and/or software) implementation plan
- Schedule printout (format is presenter's choice)
- Schedule resource usage listing (to show resource leveling)
- Copy of schedule template

The things reviewers should inspect for defects:

1. Review scheduling assumptions (Implementation Plan)
    - Are the resources defined actually available?
    - Are the resources appropriate to the task?
    - Is staff committed for entire duration?
    - Are there potential staff turnovers?
    - Is meeting time allowed for in the schedule?
2. Review the Risk Mitigation and Management Plan (Implementation Plan)
    - Are there risks missing? Look for budgetary, schedule, personnel, resource, customer, requirements uncertainty, project complexity, size and degree of structural uncertainty, and technical risks (design, implementation, interfacing, verification and maintenance problems). Also look for technical obsolescence and leading-edge technologies, business risks (sales force does not understand the product, loss of support or budget), unrealistic delivery date, lack of software scope, and poor development environment.
    - Are there probabilities and impacts assigned to each major risk identified?
3. Review all outputs and deliverables (Implementation Plan)

- Review dependencies (Implementation Plan/Schedule)
- Review risk impact (Implementation Plan)
- Compare commitments to schedule (Implementation Plan/Schedule)

4. Check resource loading (Implementation Plan/Schedule)
   - Are tasks identified at the appropriate level (project, phase, task)?
   - Are task relationships clear? Are the relationships clearly independent or sequential? Do they have multiple successors, multiple predecessors, or iterative natures, or are they just parallel?

5. Check resource leveling (resource usage listing)
   - Is this a short-term solution only? (It should be)
   - Is the trade-off between resource availability/cost and activity duration reasonable?

6. Compare the task list to template (Schedule template)
   - Is the list complete?
   - Is there any conflicting information?

CHAPTER 6

Around the World— Acknowledging Cultural Diversity

Multicultural people strive towards "totality of experience" not only by learning foreign tongues, but by cultivating empathy with the views of others, standing in their shoes in their geographical, historical, and philosophical location, seeing themselves from that location.

—Richard D. Lewis

6.1 COMING TO GRIPS WITH CULTURAL DIVERSITY

Cultural diversity is not something that is going to go away tomorrow, enabling us to plan our strategies on the assumption of mutual understanding. It is in itself a phenomenon with its own riches, the exploration of which could yield incalculable benefits for us, both in terms of wider vision and more profitable policies and activity. People of different cultures share basic concepts, but view them from different angles and perspectives, leading them to behave in a manner that we can consider irrational or even in direct contradiction of what we hold sacred. We should, nevertheless, be optimistic about cultural diversity. The behavior of people of different cultures is not capricious. There exist clear trends, sequences, and traditions. Reactions of Americans, Europeans, and Asians alike can be forecasted, usually justified, and, in the majority of cases, managed. Even in countries where political and economic change is currently rapid or sweeping (Russia, China, Hungary,

Poland, Korea, Malaysia, etc.), deeply rooted attitudes and beliefs will resist a sudden transformation of values when pressured by reformist governments or multinational conglomerates. Postperestroika Russians exhibit individual and group behavioral traits strikingly similar to those recorded in Tsarist times; these had certainly persisted, in subdued form, in the Soviet era.

International business, especially when joint ventures or prolonged negotiations are involved, is fraught with difficulties. Apart from practical and technical problems (to which solutions are often readily found), national psychology and characteristics frequently interfere at the executive level, where decisions tend to be more complex than the practical accords reached between accountants, engineers, and other technicians. Corporate cultures vary widely inside a country (compare Apple and IBM in the United States or Sony and Mitsubishi in Japan); national business styles are markedly more diverse. In a Japanese–United States joint venture, the Americans are interested mainly in profit and the Japanese in market share, so which direction do they take? When a capitalistic company from the west sets up business in a socialist country, the areas for conflict are even more obvious.

By focusing on the cultural roots of national behavior, both in society and business, we can foresee and calculate with a surprising degree of accuracy how others will react to our plans for them, and we can make certain assumptions as to how they will approach us. A working knowledge of the basic traits of other cultures (as well as our own) will minimize unpleasant surprises (culture shock), give us insights in advance, and enable us to interact successfully with countries with whom we previously had difficulty.

6.2   COMMON SENSE

When deploying processes to your global locations, you must constantly ask yourself, "Does this make common sense?" You must watch the very term "common sense" carefully, for it is not as common as it seems. The dictionary defines it as "judgment gained from experience rather than study"; the American lexicon gives it as "judgment that is sound but unsophisticated." Academics are uncomfortable with common sense, which tends to preempt their research by coming to the same conclusion months earlier. Common sense, although basic and unsophisticated, cannot be neutral. It is derived from experience, but experience is culture bound. It would seem to be common sense for the Japanese to have discarded the Chinese writing system, which does not suit their language, and takes ten years for Japanese children to learn. But they have not done so. Japan is a rather regimented society, yet the police let a man urinate against a public wall if he really has to, and will drive him home in his car if he is too drunk to drive himself. When asked why they are so lenient in such matters, they reply that it is common sense.

Whether one culture is superior to any other culture is irrelevant, and it is not the intention of this book to persuade readers that it is. What is of the utmost importance is to realize that cultural diversity does exist, and to recognize its impact and accommodate for it by simply applying common sense. To understand the language, culture, and thought process of the individuals at your global sites gives you the ability to apply common sense at each site. You will also find that being aware of cultural differences within your local team will aid in your process success.

6.3 LANGUAGE AND CULTURE

How can you set about achieving a relatively harmonious and integrated international team? To begin with, you should face the fact that to understand what makes foreign colleagues tick, there is no substitute for learning their language, reading books produced by the culture, and familiarizing yourself with the country's basic history. This requires a sizeable investment, not so much in financial terms as in time. Companies that discount the importance of such training do so at their peril. A malfunctioning joint venture with a foreign partner can result in catastrophic financial loss. Executives operating in an international framework require training, which will exempt them from the charge of complete ignorance of the culture of their colleagues. This implies language training, but also being cognizant of some very basic facts about the country concerned, such as politics, history, and geography, as well as elementary business behavior.

6.4 LANGUAGE AND THOUGHT

When training personnel on communication skills, think about the communication style of the organization as a whole and ensure that their communication mechanisms are tailored to fit how the organization works. With advocates in Europe, Asia, and India, the cultural differences will be abundant. Simply understanding the interrelationship between language and thought, as discussed in *When Cultures Collide* by Richard Lewis [21], is essential to this success. Lewis' categories, as shown in Table 6.1, illustrate the cultural differences we need to address in each communication.

Managers in linear-active cultures will demonstrate and look for technical competence, and place facts before emotion; they will be deal oriented, focusing their own attention and that of their staff on immediate achievement and results. They are orderly, stick to agendas, and inspire others with their careful planning.

Multiactive managers are more extroverted; they rely on their eloquence and ability to persuade, thereby using human force as an inspirational factor. They often complete human transactions emotionally, assigning the time it may take to develop the contract to the limit.

6.5   LINEAR ACTIVE AND MULTIACTIVE CULTURES

Multiactive people are very flexible. Linear-active people do one thing at a time, concentrate hard on that thing, and do it within a scheduled timescale. Multiactive people are not very interested in schedules or punctuality. They pretend to observe them, especially if a linear-active partner insists on doing so. They consider reality more important than manmade appointments. Multiactive people do not like to leave conversations unfinished. For them, completing a human transaction is the best way they can invest their time.

When people from a linear-active culture work together with people from a multiactive culture, it is likely that irritation will result on both sides. Unless one party adapts to the other (and they rarely do), constant crises will occur.

Table 6.1. Interrelationships between language and thought

	Description	Culture
Linear actives	Those who plan, schedule, organize, pursue action chains, and do one thing at a time.	Germans, Swiss
Multiactives	Those lively, loquacious people who do many things at once, planning their priorities not according to a time schedule, but according to the relative thrill or importance that each appointment brings with it.	Italians, Latin Americans, Indians, and Arabs
Reactives	Those cultures that prioritize courtesy and respect, listening quietly and calmly to their interlocutors, and reacting carefully to the other side's proposals.	Chinese, Japanese, and Finns

6.6 REACTIVE CULTURES

People from reactive cultures listen before they leap. They are the world's best listeners in that, as much as they concentrate on what the speaker is saying, they do not let their minds wander, and rarely, if ever, interrupt a speaker during the discourse, speech, or presentation. When it is finished, they do not reply immediately. A decent period of silence after the speaker has stopped shows respect for the weight of the remarks, which must be considered unhurriedly and with due deference.

Even when representatives of a reactive culture begin their reply, they are unlikely to voice any strong opinion immediately. A more probable tactic is for them to ask further questions about what the speaker said in order to clarify the speaker's intent and aspirations.

Reactive people are introverted, distrust a surfeit of words, and, consequently are adept at nonverbal communication. They achieve this through subtle body language, worlds apart from the excitable gestures of other cultures. Linear-active people find reactive tactics hard to fathom, since they do not fit into the linear system (question/reply, cause/effect). Multiactive people, used to extroverted behavior, find linear-active people inscrutable because the give little or no feedback.

People belonging to reactive cultures not only tolerate silences well, but regard them as a very meaningful, almost refined, part of discourse. They do not take the opinions of the other party lightly, nor do they dismiss them with a snappy or flippant retort. Finally, reactive people excel in subtle, nonverbal communication, which compensates for the absence of frequent interjections.

6.7 COMPROMISE

It is not uncommon for negotiations to enter a difficult state in which the teams get bogged down or even find themselves in deadlock. When such situations occur between nationals of one culture, there is usually a well-tried mechanism that constitutes an escape route whereby they can regain momentum without loss of face for either side. Deadlocks can be broken by, for instance, changing negotiators or venue, adjourning the session, or "repackaging" the deal. The problem remains that intelligent, meaningful compromise is only possible when one side is able to see how the other side prioritizes their goals and views the related concepts of dignity, conciliation, and reasonableness. These are culturally affected concepts and, therefore, emotion bound and prickly. However, an understanding of them, and a suitable step or reaction to accommodate them, forms the unfailing means of unblocking the impasse. Such moves are less difficult to make than one might believe. They do, however, require knowledge and understanding of the traditions, cultural characteristics, and way of thinking of the other side.

6.8   HUMOR ACROSS CULTURES

It has been said that humor crosses national boundaries with difficulty. As world trade becomes increasingly globalized, business-people meet their foreign partners more frequently and, consequently, feel that they know them better. It is only natural that when you develop a closer relationship with a stranger there is a tendency to avoid being overly serious and to begin to converse in a more relaxed manner. Swapping anecdotes is a good way of melting the ice in many situations and gaining the confidence of your listener.

Although the introduction of humor in international business talks may bring considerable gains in terms of breaking the ice, speeding up the issues, escaping from deadlock, putting your partners at ease, and winning their confidence in you as a human being, the downside risks are often just as great. What is funny for the French may be anathema to an Arab, your very best story may be utterly incomprehensible to a Chinese, and your most innocent anecdote might seriously offend a Turk. Cultural and religious differences may make it impossible for some people to laugh at the same thing. Who can say with certainty that anything is funny? If all values are relative and culture-based, then these include humor, tolerance, or even truth itself. And remember that laughter, more often than not, symbolizes embarrassment, nervousness, or possible scorn.

6.9   TEAM-BUILDING

There are a wide variety of team-building exercises and multinational corporations have tried all of them. At business schools, budding MBAs work together on hundreds of case studies. Promising managers and key staff from different countries are assembled to go camping together, climb mountains, raft down rivers, and cross deserts. A basic principle of most team-building exercises is that all members will face some kind of difficulty together and help each other out according to their individual abilities and with the resources that are at hand. The environmental constraints of a tent, raft, yacht, or classroom necessitate working closely together and avoiding needless friction. When the teams are international, interesting things occur. Individuals strive to put their personal skills at the service of the team—sometimes practical, sometimes inspirational, and sometimes intuitive. Leaders emerge—different people take charge of provisioning, planning, strategizing, financing, logistics, social affairs, and evening cooking. A language of communication evolves, as do problem-solving routines. Even in a language course, this spirit of cooperation emerges. Working with someone at close quarters for a protracted period of time enables you not only to observe foreign patterns of behavior, but to perceive some of the reasoning behind them (common sense). You also have the opportunity to explain your own actions and concepts (which are perhaps eccentric to others) as you go along. The talkative Italian, possibly irritating at first, may prove to be the social adhesive holding the group together. The disconcertingly withdrawn, opaque Japanese, sitting quietly in the corner, may later remind the group of things they have forgotten. The hustling American gets everyone to the restaurant on time, the superior Frenchman gets you the right wine, and the fussy German has a minibus and umbrellas waiting for you in the rain [21].

CHAPTER 6—SUMMARY

When working with international teams, it is very important to consider different communication styles and cultural attitudes. Remember to be sensitive to communications styles in different areas of the world that transcend individual personality types.

Address Cultural Diversity

- Common sense. Make sure your global processes pass the common-sense test.
- Language and culture. Learn about the cultures with which your company is working
- Language and thought. Understand the interrelationships
- Linear-active and multiactive cultures. Schedule and detail focus versus flexibility
- Reactive cultures. These are the "listen first" cultures.
- Compromise. Use your understanding of your culture types to manage negotiations for success.
- Humor across cultures. Be sensitive about how one person's humor can be another person's insult.
- Employ team-building exercises

CHAPTER 6 AIDS—"PEARLS OF WISDOM"

This exercise requires a team facilitator and either whiteboards or large poster paper.

Gather your team (up to 8 people) in a large group and explain the purpose of this exercise is to share pearls of wisdom. If you have a larger group, split them into teams of 5 or 6 apiece. Ask each person to write down a pearl of wisdom that they remember being passed along in their family. This could be either a saying from their parents or a famous quote that they heard repeatedly. Give them a few examples like: "That dog don't hunt" or "All that glimmers is not gold."

Allow for 10 minutes of brainstorming time and have each team write down their sayings either on a whiteboard or poster sheets—large enough for the whole room to view. Ask each person present to the group at large. If the group is too large, have people present in their smaller groups.

Instead of requesting the person who wrote each pearl explain it, have a different person from the group read the saying and interpret it. Create some fun with this and encourage participation from the team at large. Answer the questions: What does the pearl mean? When could you use this pearl? Is it is helpful or hurtful?

When everyone has contributed to their version of the meaning of the pearl, ask the originator of the pearl to explain: What the saying meant, when it was usually said, if there are any implications of that saying for them in their work life, and finally if they still use the saying. If so, when and why?

Rotate through as many pearls as you have time for then summarize the 5 most useful or interesting phrases for the workplace and post them on the whiteboard/poster.

Have the team discuss any particular nuances that result from colloquial terms. Open the discussion to how people have phrased words that have been confusing or disconcerting. In the combined group discuss how these sayings can apply at work.

You can use the final 5 pearls to create small posters or insert them in presentations to remind everyone of the need to be sensitive to how phrases are perceived. This can also be used as a reminder of the team building event and create a feeling of camaraderie.

CHAPTER 7

Move Your World— Managing Change

If you want to build a ship, don't drum up people together to collect wood and don't assign them tasks and work, but rather teach them to long for the endless immensity of the sea.

—Antoine de Saint-Exupery

7.1 INSTITUTIONALIZING CHANGE

Your organization will need to have a basic change-management mechanism in place in order to "institutionalize" your changes. Now, the word "institutionalize" has always bothered us. Images of automatons, forced, to wear white straight jackets and walk the white halls of a pristinely clean insane asylum, come to mind. These automatons then go blindly go about their days, oblivious of their surroundings. These are the sorts of images that word inspires! This would definitely not define our style of continuous process improvement! Institutionalization refers to embedding processes so deeply into organizations that those processes become "just the way we do business." We believe change should be messy, dynamic, and colorful, yet we still promote basic change-management processes to assist in the creation of permanent, constant change.

Just as you would plan a project, so should you plan how you will manage change. You will need a basic management plan of who, what, when, where, and why, along with the communication pathways that you need to get the word out. We have already discussed chartering the targeted changes and communicating that out; however, most of the angst about change in a company is that the change is undercommunicated and does not include the individuals in the organization. Because we are talking about a continuous change process, you can build an evergreen management plan that includes guidelines on how to manage ongoing change. By documenting the CPI effort in your organization, you will be almost there. You have already garnered management sponsorship for your change efforts, chartered each effort, and engaged with champions and advocates as well as the rest of the affected organization. However, there is the large component of communications that you must address.

7.2   COMMUNICATIONS

Your communications should include information on why your organization is changing in any particular area. People want to know the compelling reason for change, along with why it will be valuable to the organization. This will help build the always-needed buy-in for change. The CPI process by nature includes simultaneous initiatives, and must be nimble enough to be fast paced. You will need to develop a common understanding of why your organization requires constant change, and you should compel it to engage in the fun.

Part of compelling your organization is building a sense of urgency and a feeling that the effort will succeed because of the quick wins that you have already demonstrated. Building these pictures for your organization will be important in attaining not only buy-in, but also excitement about the ongoing efforts. Your communications plan will be key in spreading the word about what is going on so that each individual in the organization gets a sense of success and excitement about what is going on.

Employees will want to know how changes will affect them and what the consequences will be if they do not change. They will want to understand What's in it For Me (WIIFM). Oftentimes, individuals see change as an exercise far removed from their everyday work. CPI principles pull in many individuals but cannot touch them all, so you need an organized method for communicating needed information out to everyone at all levels of the organization.

When building your communication plan, you need to identify the goals of the communication. You are likely to have multiple communication goals that you will want to raise awareness about to your employees. Consider your organization's breakdown and understand who your audience is and what level of information they will want to hear. List groups as well as job roles, work responsibilities, and communications paths. You will want to build your communication pathways and understand them so that you can communicate effectively and distribute information at the right levels. Whatever outline for a communications plan you use, you will want to consider delivering the message in a combination of facts and data with ideas and concepts. In order to be effective, you must communicate in many different directions with different types of messages. You can combine graphics, text, presentations, e-mails, and newsletters together to make a complete communication picture.

7.3 BARRIERS TO ADOPTION

It will also be important to consider, identify, and address any barriers to adoption that may exist in the organization. Your sponsor should be able to help navigate these organizational barriers and clear the way for you. Another barrier could be organizational culture. Perhaps your organization is particularly resistant to change because they have seen many initiatives come and go with no results. This may have resulted in a sense of apathy and resistance to change, because there is no consequence for not changing.

There may be a general culture of apathy or resistance that is a company inheritance. Depending on how young or old your organization is, you may have issues on both ends of the spectrum. A young organization may not see the need for any sort of process or evolving process because, since it is new, it expects total chaos. Older, more established organizational cultures may perceive change as unneeded altogether because they have done well so far and think that the status quo is good enough. Either way, to overcome these barriers to adoption, an organization must build both a sense of urgency and a sense that constant change and adaptation are important for survival.

7.4 PILOTING

As explained earlier, it is imperative that you demonstrate some quick wins to the organization in order to build confidence about the need for change. By demonstrating how the process improvement will improve not only the business but each individual's work environment, you will prove that this is not a flavor-of-the-week effort. Piloting can help you evaluate whether or not you will obtain the return on the change that you expected, will set up the organization for successful launch, and will help you refine the process that you are changing. This makes it palatable to the stakeholders involved.

In your early efforts, you will want to pilot new changes in small areas to test them out before you expand the changes to the organization at large. Select an area that you can monitor and measure, and then ensure that both the communications about that effort and the effort itself is successful. These small pilot efforts will teach you much about the larger effort, and can save you from perceived failure.

It will be very important that you select appropriate pilot projects. Do not pick a weak or failing project, but rather select one that is relatively strong and has enough time left in it to be able to demonstrate what you are trying to prove in the pilot. You can then extrapolate results from that test, and, along with lessons learned, apply what you learned to the organization at large.

Piloting is especially important if you are charting new waters, particularly if you are using new technologies or new methodologies that the organization has not used in the past. This will also give you a sense of whether or not it will require additional training before you implement a change. You will also find out if you have a terminology problem in the organization and if you can correct this early on. Common terminology, or organizational vocabulary, can be a very large issue if not addressed early in a project. What is an orange to one person could be a mandarin to another, and this can lead to confusion and frustration.

7.5 IMPLEMENTING

Of course, any change management (or accelerated change) effort must get to the point of implementation. Introduction and communication of the change is going to be of great importance to its final success. The CPI effort itself could be considered a change effort, because it requires that the elements of change be built into existing processes.

You must drive adoption with the idea that senior management endorses the change. The best person to communicate that implementation has started is the CEO. When the CEO puts forth announcements regarding the change, individuals usually decide that the effort must carry great weight with him.

Although we like to say that you cannot directly motivate an individual—not for long anyway—you can provide an environment in which the individual feels motivated. Implementation is the stage at which you will determine if your individuals are motivated. If you are mandating change, it is particularly important to describe under what situations you must start the new implementation. You should answer the following questions for your organization: Is the change going to occur.

- Immediately?
- At your next phase?
- After you have completed the project you are working on now?
- Only after you have performed training?

Be sure to consider any other questions about the implementation. Will users need to migrate from some other system? What about users who are not co-located? Will they all migrate to the new process simultaneously, or will there be a sequential rollout? This information should be included in the communication plan.

During implementation, you will also want to track results and report them at various levels. Be careful that the pace of changes made are not so dizzying that the organization cannot keep up. Allow people time to buy into the changes before they need to implement a new set. All changes made need to fit in with the organization's overall culture and climate for change.

7.6 LESSONS LEARNED

Keep track of missteps along the way and ensure that you avoid making the same mistake twice. You will want to track these lessons throughout the project and in the life of other projects so that you will more effectively roll out your process improvements. Sometimes, it is difficult to stop and capture a lesson learned in the midst of "doing the work." You need to devise a way—perhaps by using a simple spreadsheet—to capture lessons learned for future analysis. It is good to put them in an electronic format for easy archiving and distribution.

In order to be learned, a lesson should be clear, concise, and in context. A clear name for the lesson should be selected, along with what it impacts and how the issue arose. You should catalog and eventually group this with other similar lessons. This need not be a heavy process, but it is a necessary one. By reviewing lessons learned and capturing them for future use, you will be able to improve how you evaluate your initiatives.

CHAPTER 7 SUMMARY

We have defined several components of change management throughout this book. This chapter emphasized the need for change management and the components required to complete the picture. These components are:

- **Institutionalizing Changes.** Balance the need to implement widespread change that is repeatable throughout the organization with the need for being agile in process change.
- **Communications** should include what the message is and who your target audiences are, as well as the pathways by which you will communicate. An outline of a communication plan is included as a chapter aid in the next section.
- **Barriers to Adoption.** Be aware of barriers to adoption such as lack of buy-in, insufficient communication, confusion, and a company culture of resistance and apathy.
- **Piloting,** Think about how you can pilot an initiative to learn about how well it fits in your organization. This allows for corrections before you attempt to roll out a process to the organization at large.
- **Implementation.** Determine how you will implement the changes. This implementation plan should include the time frame and which projects will roll out the new process. Also consider all operational locations and when they will adopt the change.
- **Lessons Learned.** Ensure that you have a method and electronic capture system for your lessons learned. Make sure they are clear, concise, and can be acted on. Consider grouping them in logical categories for future reference and analysis.

CHAPTER 7 AIDS—COMMUNICATION PLAN

Use the following outline to build a communication plan. This checklist is designed to assist you in working through the following components of your plan: strategy, target audience, communication goals and objectives, and communication pathways.

- **Initiative Description**—what is the issue that is being addressed by this communication plan?

- **Target Audience**—list each target group that you will be communicating to and the purpose of the communication. Be sure to include any special remarks such as level of communication and groups.
- **Target Locations**—list out any locations that you will need to communicate to and any special needs due to the location.
- **Communication Goals**—define what measurable, defined actions you will take.
- **Communication Objectives**—define what intermediate steps are required to meet the communication goals.
- **Communication Paths**—define how you will reach your target audience and by what means.
- **Messaging** —for each group with whom you will communicate, define:
  - **Message**—what message you are trying to convey
  - **Method**—activities, materials, and communication paths, such as e-mails, presentations, and on-demand video.
  - **Frequency**—how often and on what time line you will communicate to each group
  - **Resources and Budget**—define any resources and budget needed. Include any personnel needed.
  - **Communications Matrix**—a consolidated view of the communication plan by message:

Message	Purpose	Person Responsible	Target Audience	Timing
FAQs	Provide answers to most commonly asked questions	Quality Director	Everyone	Ongoing and updated monthly
Online Newsletter	Disseminate information	VP Engineering	Everyone	Quarterly
. . .				

- **Measures of Success**—what will define success for the communications plan?

CHAPTER 8

Rock Your World— Encouraging Process Perpetual Motion

Every morning in Africa, a gazelle wakes up knowing it must run faster than the fastest lion or be killed. Every morning a lion awakens knowing it must outrun the slowest gazelle or starve to death. It doesn't matter if you are a lion or a gazelle; when the sun comes up, you'd better be running.

—Martis Jones [22]

8.1 CONTINUOUS IMPROVEMENTS

The key to continuous improvement is that it is, well, continuous! The process is never completed. You are never finished. Because your products and services must change over time, your processes and checklists to support them should also evolve [23]. Quality is in the eye of the customer, and as your customer changes, your Quality will likely change. You need to be careful not to fall back into rigid, unchanging processes. As soon as you are certain you have decided on the most effective process and the method to document it, someone will find a way to improve it. This continuous improvement is a good thing, albeit change is usually uncomfortable. Every aspect of ensuring Quality is an evolutionary process. You will want to ensure that you continually use these tactics to keep the organization moving in the right direction.

In the book *The Blind Men* and *the Elephant: Mastering Project Work* [24], David Schmaltz states it this way: "We create maps

Collaborative Process Improvement. By C. L. Yeakley and J. D. Fiebrich
Copyright © 2007 IEEE Computer Society.

Here's a good example. As a lead assessors, we were often called upon to assess an organization's software capability. In great anticipation, a team would look forward to being assessed at a particular level (according to Software Engineering Institute guidelines). An organization must learn to focus on the end game instead of focusing on the singular achievement. Thus, if your goal is to be able to run a marathon, you first have to walk (level 2). But because your goal is to be able to run (level 5), in preparation, you may have to incorporate some small sprints into your walking routine. It is the same with process improvements, particularly for software. You may be dubbed a great walker (level 2), but if you did not prepare to run (level 5) in the first place, you may be stuck being a great walker for your entire career. This is because you did not plan on the next step. If you had, you would have challenged yourself to do a small bit of running so that you could prepare for the next step. If you do not mix in some small runs, you will not know what it will take to win that marathon.

You can't plan everything up front. If you could, there would be no need to continuously improve, you would not have to build prototypes to test systems, and you would not have to change your plans when the unexpected came your way. How many times has it happened in your own life that when you had to make a substitution in a recipe (project plan), the final product turned out better than if you had the original ingredient on hand? Sometimes, realization of adversity will result in an improvement. When you not only anticipate the unexpected, but prompt the unexpected to happen, then you can truly move your processes to the next level. After all, if you do what you have always done, you will get what you always got. This is even the case in the improvement business; you have to keep moving in order to prevent satisfaction and success from ruining your future.

without surveying the territory." This is a statement that we can take to heart when talking about CPI. We do have to create a map of unknown territory. Perhaps the edges are even fuzzy and we try to tidy them up the best we know how with the information at hand. The truth is that projects, like life, are messy, and CPI is no exception. In order to keep the overall process alive, you need a little dynamic tension, something to keep the idea of continuous improvement moving forward.

One way to keep process engagements from becoming stale is to keep looking at your processes in lateral and unique ways. A less prescriptive approach for helping members of organizations identify and carry out improvement opportunities is to provide a framework for looking at the organization in new ways. In Peter Senge's book titled *The Fifth Discipline: The Art and Practice of the Learning Organization* [25], the learning organization is such an example, emphasizing the need to find ways to embed continual transformational learning into key business and personal processes and activities.

Methods to help your group cope with the evolution include the following:

- Using diverse, cross-functional team members
- Executing continuous process improvements
- Migrating activities into an automated tool

8.1.1 Using Diverse User Groups

A crucial ingredient in process improvement is diversification. To have marketing, engineering, and test conduct a round-robin discussion concerning configuration management is an eye-opening experience! When developers, testers, and managers brainstorm a new process for requirements management, you can feel the electricity in the air. Before the meeting is over, each individual is able to think about the issue in a completely new way. Cross-pollination of ideas is an invigorating experience. At your company, you will observe that individuals will continue discussions long after the meeting is over, and will always arrive fully prepared for the next meeting.

The question now becomes "how do you diversify?" You will need to engage your management to get true diversity and engagement across the organization. If you do not do this, then your processes will not take hold and your improvement efforts are likely to fail. It is imperative to remember that foremost in employee's minds is "What is in it for me?" If you cannot answer that question

Side Effect. Having diverse user group participation was comparable to having a peer review of a project schedule at the project kick-off meeting—an eye-opening experience for all involved.

sufficiently, the organization will not embrace any change, even if it nets the company a cool $50 million in savings. If the employees do not realize this gain, then they will not be motivated to change.

Say you are on the software development team and you are trying to get improved requirements. You start performing formal reviews of requirements, but marketing is not participating. They do not care about reviews; they just want the developer to deliver a product that their customer wants. They want the developer to understand the customer and deliver on the mark according to what they specified.

The problem with this approach is that the developer may not understand what the marketer really meant when they said "improved performance." The marketer may not understand what percent improvement is and now you have a set of requirements that only one side understands.

At first, you may meet with resistance when trying to engage the marketing (and other) team in your process improvements. Individuals tend to not be able to visualize someone else's work until they see something tangible that they can understand. It may take the effort of locking up cross-functional teams up together in a room to make this happen. The way we did it was to perform a live technical review of a product specification with a cross-functional team. It was only then that "light bulbs started turning on" across the functional teams.

It's not until you can *show* the teams what the problems are that they will all understand the problem. In order to do this, you have to figure out a way to get that cross-functional communication going and engage everyone in the process, whether or not they see the value in it at first. It is your job to show the entire organization the overlaps and dependencies. You cannot work in a vacuum and get optimal performance.

A simple approach is to engage management in your quest. Get your management sponsor to talk to his or her peers and get them to engage in the process improvement effort wherever it makes sense. Refer to the Table 8.1 for a matrix of possibilities and areas that use the Software Engineering Institute Key Process Area for Software as an example.

Table 8.1. Support interfaces

Process Area	Software Engineering	Marketing	Test Engineering	Operations
Requirements Management	X	X	X	X
Configuration Management	X		X	
Planning and Oversight	X	X	X	X
Project Tracking	X		X	
Software Quality Assurance	X		X	

Think about your organization and your identified process areas, and make up your own matrix to determine your overlaps. Wherever they are, you will need to build a business case that explains why you need engagement from that functional area. Then you will need to ensure that your business sponsor can recruit his or her peers and engage them in that process area. Recruiting does not mean a halfhearted attempt at pushing the new flavor of the day; it means, "This will appear on your performance review and will affect your raise next year." The effort must count and the team members involved must be rewarded for their efforts.

When you are sure that you have identified all the key personnel, including functional areas that may be affected, then you are ready to begin building checklists that will serve as helpers in tracking your progress in implementing your processes.

8.1.2   Executing Continuous Process Improvements

When questions are raised like, "When will this process be done? How much more work on this process is needed?" the role of a great Quality Manager is to say, "We will be constantly improving all of our processes." At a time when your projects are heavily driven by milestones and schedules, it is difficult to keep individuals motivated when there is no plan for the effort to be finalized [26].

When establishing and maintaining processes that improve overall quality, you will find that the effort is never finalized. Just as you strive to improve your products every day, you must strive to improve the processes that create them every day. At your company, mitigate these reservations by engaging in metrics gathering. Aggressively recruit individuals to participate in complementary Six Sigma efforts. The principles used in this define/measure/analyze/improve/control (DMAIC) method not only aid advocates in active, measurable process-improvement activities, they also support measurement activities for your designated standard.

You can help encourage active engagement in CPI by publishing process helpers on your local intranet. Ideally, you would have a dedicated Uniform Resource Locator address that personnel could easily remember and refer to. You can also use this site to advertise, if you will, the current activities and state of the practice. In this way, you actively put process information, including templates and metrics, at their fingertips, which will likely spark more innovations.

The diffusion of an idea is a social process, not one you can legislate. The success of any innovation involves early adopters, the majority, and laggards. The rate at which employees adopt any change can be adjusted in several key ways. Each can raise or lower the stickiness (how well it takes hold) of the change through the following variables:

- **Relative Advantage.** The greater the perceived advantage, the more rapidly it will be adopted.
- **Compatibility.** How compatible is it with what you already know?
- **Complexity.** How difficult is it to learn and use?
- **"Trialability."** How easy is it to try?
- **"Observability."** This is the degree to which results are visible.

Although these five stickiness variables make up 49 to 87% of the rate of adoptions, there are a few other variables involved: the type of decision (top-down or word of mouth), network, social system, and change agent's promotion. Top-down is a faster way of starting the change. Word of mouth is the next fastest way and minimizes the risk of rejection.

By using the CPI guidelines we are giving you in this book, you can engage at both the top-down level and the grass-roots level. This will be key in accelerating change for the better in your group or organization. CPI guidelines require that you not only engage top-level management, but also include middle managers as sponsors and individuals as daily collaborators.

The group you will create and empower will be a daily, living microcosm of your organization as a whole. It is important to be able to take these improvement activities and build them into a system with which you can track and monitor the activities themselves so that even they can be improved.

8.1.3   Migrating Activities into an Automated Tool

Coordinating and orchestrating advocate activities is a full-time job. You can use spreadsheets to schedule tasks, document compliance, and initiate reports, but they are probably the most difficult tools for accomplishing this task. Using a powerful, automated program management tool enables you to remove all of the personality from the equation. When you schedule assessments and reviews as project tasks, they will be conducted systematically. This will standardize software quality engineering activities for all projects. Having the tool automatically generate progress reports on a daily, weekly, or monthly basis will become a matter of routine.

Any time a tool is introduced, there is a tendency to make the process fit the tool. Remember to focus on the process, not the tool. The tool does not yield the results; the process and people do. The best tool can be quickly undermined if it is not supporting the process. Understand the implied process of the tool and assure that the tool can stretch to cover the gaps. For example, a waterfall tool does not support iterative development well.

CHAPTER 8 SUMMARY

Continuous Improvements

- You are never finished
- Keep the organization moving
- Continue to push the organization to learn

Using Diverse User Groups

- Integrate users from cross-functional areas for input and cross-pollination
- Be prepared to answer the "What's in it for me" question
- Get support for cross functional interfaces by engaging management

Executing Continuous Process Improvements

- Aggressively recruit high-performing personnel
- Link to Six Sigma efforts
- Pay attention to stickiness factors: relative advantage, compatibility, complexity, trialability, and observability
- Actively involve all levels in the change

Migrating Activities into an Automated Tool

- Utilize an industry standard tool to manage CPI tasks
- Ensure that your tool matches your process
- Ensure that the level of extensions of the tool is not so large as to mitigate the annual or biannual releases of the tool from the vendor. Allow improvement from vendor sources to your tool's support.

CHAPTER 8 AIDS—SAMPLE ORGANIZATIONAL CHANGE

Readiness Checklist

The following list could be posted in a process newsletter, or sent out as a survey to the workers and/or improvement team to help in determining the readiness of your organization for change. If the preponderance of answers you get to this checklist is negative, then you should implement more work in communicating to appropriate levels in the organization.

Commitment

1. Our President/Chief Executive Officer fully supports this change.
2. The management team will include change processes in rewards and recognition programs, including yearly performance evaluations.
3. Consider advocates and champions for high-visibility positions in the company, including opportunities for promotion.
4. All managers must understand the change in culture that will be needed for CPI and the changes that will result.

5. All levels of management must consider CPI and the resulting changes as important to the future of our organization.
6. The right thought leaders are engaged in this change effort.
7. You can align CPI efforts with organizational goals. The charter reflects opportunities to support short- and long-term strategic goals of the organization.

Infrastructure

1. Proper communication pathways are available to inform the organization of these changes.
2. Information technology support for this project is available.
3. You can successfully implement this change within the current organizational structure.
4. Human Resources are engaged and have the ability to create an atmosphere in which people will be recognized and rewarded for being advocates and champions.
5. Workers will be appropriately compensated for engaging in CPI efforts and will be evaluated for a proper attitude to change.

Advocates and Champions

1. Are allocated enough time to focus on their key areas of representation.
2. Are (or there are plans in place to train them) properly trained and coached to perform their duties in the CPI process.
3. Will be considered for promotions at a rate higher than the general population.
4. Have full management support and visibility.
5. Champions include thought leaders of the organization.

Culture

1. We have considered all cultural barriers to success and evaluated our organization in terms of communication styles.
2. We have considered opportunities for tailoring to regional team's operations if there are geographically dispersed teams.

CHAPTER 8 AIDS—INTERNAL WEBSITE

The following sample web pages are examples of what information you could post on internal websites. This is another good way to communicate information about process best practices. Of course, the Web pages need to be easy to find and read, so you should review these web pages just as you would review any other documents. It is preferable to link the pages together and make them easy to navigate, as well as hold all the pertinent information and contact information. In this way, it would be possible to use them as a recruiting mechanism. If the project looks fun and rewarding, then more people will want to be a part of the overall process.

CHAPTER 8 AIDS—BEST PRACTICES WEB PAGE

CONTACTS

Process Team

Forum

best practices

Improving Project Management, Engineering, and Software Development Processes

PROCESS

Policies

Product Development Process (PDP)

Templates

o   By Phase

o   By Role

Project Notebooks

Reviews

Metrics

TEAMS

Process Team

Development Methodology (DM)

Pubs

TOOLS

Team Play

View CVS

company goals 2007

1. Execute flawlessly on our commitments.

2. Exceed $60M external revenue.

3. Define and start developing the differentiating hardware, software, and services within our target and vertical markets.

resources

Getting Started with Process

o   For an overview of the process of how we create products, read the Product Development Process (PDP) pages.

o   To learn how Software Engineering Institute methodologies help us create products of quality on time.

o   Need to learn about specific templates and when to use them see the Templates page.

Training

o   Weekly Training Hot Sheet

Help | Glossary | Feedback
Last updated April 2006

CHAPTER 8 AIDS—POLICIES

Policies

Corporate policies on process and quality are available below.

Policies	Description	Other	
Configuration Mgmt	Software Configuration Management (SCM) effectively manages the evolution of software throughout all phases of development through obsolescence.	Owner Author Reviews Authorizes Last Rev Next Rev	
Project Planning	Software Project Planning (SPP) establishes the project scope and objectives and to identify and define the resources required to successfully deliver quality products and services to the project sponsor.	Owner Author Reviews Authorizes Last Rev Next Rev	
Project Tracking & Oversight	Project Tracking & Oversight (PTO) provides adequate visibility into actual progress so that management can take effective actions when performance deviates significantly from the plan.	Owner Author Reviews Authorizes Last Rev Next Rev	
Quality Assurance	Software Quality Assurance (SQA) policy provides visibility into the quality aspects of the products being built, and the processes being utilized by the Software Development units.	Owner Author Reviews Authorizes Last Rev Next Rev	

Help | Glossary | Feedback   *Last updated April 2006*

CHAPTER 8 AIDS—DEVELOPMENT METHODOLOGY TEAM

development methodology (dm) team
"If you do not change the system, nothing changes."
-- Stephen George

latest news
Current initiatives (date)

Scarab is being updated with reporting improvements and choice user comments

Replacement to CVS under examination (subversion)

Bound checker testing underway on three different projects

Improvements to the Formal Technical Review process via automation and web interaction are in the works

Best Practices web pages under review and reconstruction

Previous initiatives

FTR metrics being collected and collated for a "state of Formal Technical Review" evaluation

resources
meeting information

Time: Every Friday, 1:30PM CST

Location: Texas Conference Room

Dial-in: 1.888.555.8686

Intn'l: 1.303.555.3287

Conference ID: 155536177

History: Meeting Notes

Help | Glossary | Feedback *Last updated April 2006*

CHAPTER 8 AIDS—PRODUCT DEVELOPMENT PROCESS

product development process
"If you can't describe what you are doing as a process, you don't know what you are doing."
- W. Edwards Deming (1900-1993)

planning phase

Purpose *Defines the project plan and commits to all downstream milestone dates and completes Sign-off.*

Inputs	Phase	Outputs
Documents An ***approved*** Software Project Management Plan (SPMP)	*PLANNING*	**Documents** Reliability Test Matrix (RTM) High-Level Design Documents (HLDD) Low-Level Design Documents (LLDD) Software Test Management Plan (STMP) Product Documentation Estimate (PDE) Software Project Plan (SPP) TeamPlay schedule

Steps to Achieve

Based on approved Work Products from Definition Phase

> Developers begin writing design specs (HLDDs and LLDDs)

> Test lead begins writing STMP, including update to TM

> Doc writer begins writing PDE

> Project Manager begin writing SPP

Reviews of the above Work Products are conducted during generation

All design specs are finalized and completed

The STMP and PDE are finalized and reviewed

The TeamPlay schedule is finalized

The SPP is finalized

The SPP and TeamPlay schedules are reviewed for approval

The MSO baseline is set in TeamPlay, and the SPP is posted as final

---

Help | Glossary | Feedback
Last updated April 2006

CHAPTER 9

Your World of Influence— Sneezing and Spreading the Improvement Virus

Management is nothing more than motivating other people.
—Lee Iacocca

9.1 SNEEZERS

The Quality Manager must put on his or her best consultant hat and step out of the limelight to encourage a non-Quality person to step up and be an example for the group. In *Unleashing the Ideavirus* [27], Seth Godin calls opinion leaders "sneezers," people who infect the population by sneezing the idea to other people. These opinion leaders can make or break an implementation. In essence, you may have the right idea—get commitment of the leaders—but you could focus on the wrong leaders: the formal ones, not the informal ones. Just as the leader of a gorilla tribe would be an awesome, fearsome, respected silverback, the thought leader could be the 500-pound gorilla who could garner support without a fight. Even though you do not want to encourage this behavior, you can use your thought leader to blaze the trail for these methods. If you cannot locate someone within the project group who is willing to dedicate time to this effort, you may have to recruit a new, independent eye from outside the project group. Without question, it is essential to have someone well known and highly respected play this important and influential role and lead this activity.

Managing by influence employs the following [28]:

- Using champions
- Engaging upper management
- Ensuring annual goal commitments
- Obtaining executive team participation

9.1.1 Using Champions

Having a champion for software quality engineering activities is an absolute for communication. This person is responsible for recruiting and training members of the Advocate Program. He or she should document, brief, evaluate, and improve the training curriculum.

During the first 12 months of our Advocate Program, we saw the enrollment increase from 3 to 30, as shown in Figure 9.1. That is a 1000% increase in participation. This growth was directly attributed to the communicated importance from upper management and the positive project impacts of the program that were the subject of many operational review meetings.

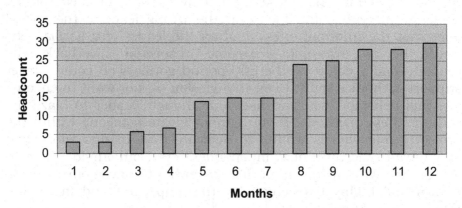

Figure 9.1. Advocate population growth during the first year.

Having the entire training curriculum available for use and review is helpful. Having a champion as the point of contact for all activities is even more helpful. The champion is often:

- The liaison between the advocates and management
- The essential source for problem resolution
- The initiator of new and improved processes

Champions of each key process area are also responsible for process documentation, improvement, and training. They can essentially function in the same manner as a software engineering process group.

9.1.2 Engaging Upper Management

As Robert Hayes and Steven Wheelwright state in their book titled *Restoring Our Competitive Edge* [29], productivity dominates the concerns of today's executive management. Stephen George goes further in "How to Speak the Language of Senior Management" [30] by saying that if the field of Quality is to get the attention of top management, then it must learn to integrate quality initiatives with financial performance. It is for this reason that we recommend that at least one aspect of the process change have tangible, measurable success factors associated with it.

In our organization, we initiated a green belt, Six Sigma effort on formal technical review metrics. We were able to tally up specific dollar amounts associated with the cost of poor quality and report this to upper management.

Face-to-face reporting of the advocates and champions can accomplish this, since they will be engaging upper management on a regular basis. Moreover, the advocates and champions must be the ones to initiate this action. Upper management has many responsibilities, of which advocate and champion interfaces are parts.

Once engaged, members of upper management will voice their support and expect the advocates and champions to continue their successful execution. It is up to the advocates and champions to go back to upper management as often as necessary. They should brief upper management on successes as well as setbacks. If not approached again, management may believe that all is well and successful.

One key tip for the Quality Manager: Never assume that upper management "gets it" just because you have explained it to them. Advocates and champions should plan on monthly status updates and formal reports. It is also the responsibility of the champions and advocates to share information from management meetings with other champions and advocates.

There is a great danger in not pursuing continued improvements that do not have hard costs associated with them. This is often the big push in larger businesses: "Finance wants only *hard* costs," they will not count soft costs of saved time. Unless members of their team are engaged in a CPI effort, finance will not be able to attribute improved quality of the final product to the process that got to that level of quality. If they are engaged, not only will they see the return on investment for soft costs, but they will also help the champions and advocates quantify them.

9.1.3 Ensuring Annual Goal Commitments

Because financial figures reflect only one perspective of business, Robert S. Kaplan and David P. Norton, authors of *The Balanced Scorecard* [31], proposed a balanced scorecard that integrates the scorecard with managing business strategy.

The elements measured and presented in the scorecard represent the best elements needed to present a balanced view of the business. For example:

- Financial factors
- Customer factors
- Internal factors
- Organizational learning factors

In the book, *Performance Scorecards: Measuring the Right Things in the Real World* [32], Richard Y. Chang and Mark W. Morgan present a measurement process they call "performance scorecards." Clearly, scorecards, sometime called dashboards, recognize the need for making decisions based on a balanced perspective of the business. This only serves to help the case for CPI.

By being able to see scorecard factors as being important in the whole, you can raise that perceived expectation for process improvements.

> In our particular case (in software), our balanced scorecard actually had our standard for software process maturity called out as a goal. This supported customer, internal, and organizational learning factors. We tied much of the efforts we drove by referring to this goal on the balanced scorecard.

If your organization does not use the balanced scorecard method, you can consider other tracking methodologies. In the book, *Managing by Measuring: How to Improve Your Organization's Performance Through Effective Benchmarking* [33], Mark T. Czarnecki reviews a number of measurement processes and programs. This led to the concept of benchmarking measurement programs with other businesses, and ultimately to reengineering an organization's programs for greater effectiveness. Whatever way in which your organization works, be sure tie in your CPI efforts with the resident measurement program.

By using a vehicle such as a balanced scorecard to communicate annual goals, you can ensure that testers, engineers, and managers are aware of and support the advocate effort. It is wise to make the goals of the advocate program as visible and equally as important as the engineering goals. The rewards associated with engineering goals should map to the rewards associated with advocate goals. Communicate that the achievement of Advocate goals is essential to the success of the company and that these goals also deserve executive attention.

9.1.4 Obtaining Executive Team Participation

Having executive team members participate in activities is essential, even if executive team members only attend team briefings. Having the executive team meet with the advocates and repeatedly reassure the advocates that they are empowered to stop the project if they find it to be noncompliant is important to the success of this effort. It is also important for advocates and champions to attend regular meetings with upper management on at least a monthly basis.

Making advocates and champions visible in the executive environment is as important as having executives visible in the advocate environment. Better yet, if you have the ability to fast track a qualified advocate or champion to more responsibility, the organization at large will take notice, and this will cause individuals to actively pursue one of those positions, as it will be seen as highly regarded by management and a path to career success.

9.2 DEALING WITH DIFFICULT PEOPLE

As professionals, you have been asked at one time or another why you enjoy your sometimes-crazy line of business. It is certainly much easier to manage rental and commercial office space than manage a team of engineers and designers, if for no other reason than, in rental, if a tenant does not wish to abide by the rules, they can be asked to leave. You do not necessarily have to influence them to do it your way. It is easy to manage by authority but much more difficult to manage by influence. The results, however, are more effective when managing by influence. To effectively deal with difficult people, consider:

1. Why are people difficult?
2. How do you determine what is really bothering the person?
3. What are the practical methods of handling difficult people?

9.2.1 Why Are People Difficult?

There are many reasons why people are difficult at one time or another. Some people are naturally disagreeable and have never been able to adjust from their behavior. You cannot pretend to interact effectively with these people because their attitude prevents you from dealing with them in an effective manner. However, many people experience difficult times due to variety of valid reasons, while at other times, the reasons may seem petty. The underlying reasons for an unreasonably demanding attitude may be related to job stress, home stress, traffic, lack of understanding, you-versus-them syndrome, having a hidden agenda, and even poor service.

From the perspective of the Board of Directors, a lack of demonstrated leadership can cause and encourage confusion, conflict, and intolerance. On the lighter side, but still applicable, blame can be placed on the weather, lunar movements, football losses, or having teenage children! As a leader, being aware of these human frailties in yourself as well as others will go a long way in dealing with the individual who is being difficult.

9.2.2 Finding Out What Is Really Bothering Someone

The most difficult part of working with this issue is trying to identify what the problem is so that you can find ways to solve it. Each individual is different, and having the patience to delve into the problem is an area to focus on in order to be more effective. The first requirement is to be a good listener. A difficult person is often someone who wants attention. This does not necessarily mean they do not have a valid issue. However, their demanding approach discourages others from wanting to explore the issue further so that they can provide assistance to them. Being a good listener requires patience, focus, and understanding. Lacking these qualities could cause failure in determining what is genuinely bothering someone. If you are successful in being a good listener, you can often use your influence to create a calming climate, and eventually have a nonemotional discussion. Encouraging individuals to discuss their issues without being condescending or judgmental will enhance your ability to narrow down the issues and subsequently find a solution. When an individual believes you are sincerely trying to solve his or her problem, that person will respond positively, even though the solution may not have been what they expected.

9.2.3 Practical Methods of Handling Difficult People

As leaders, you know that when conducting meetings you can use (or hide behind, if you wish) Robert's Rules of Order and instill a certain amount of control the legal way. This may not always be in the best interest of the group you are representing, and, in fact, can destroy the relationship you have nurtured over time. When evaluating practical methods of handling difficult people, the first and foremost thing to consider is what is right for a particular situation. If you lose integrity with a difficult person, the next time you deal with a difficult person will prove to be two or three times harder. Maintaining flexibility and having an open mind on both problem solving and positive thinking is extremely important. People respond favorably and with a cooperative attitude when you offer a positive perspective while giving them the enthusiastic impression of "Yes, we can solve this together." When individuals are difficult and present you with a problem, ask them, "Can you recommended a solution?" Whenever possible, make the person a part of the solution; influence them to "own" the solution and to take credit for it.

Often, you are dealing with difficult people within your company. This usually occurs at meetings. Planning a well-thought-out agenda with appropriate structure provides the proper vehicle for a successful meeting. Many times, if you conduct meetings with a good structure, it is easier to deal with difficult people. Having periodic meetings with a formal agenda will also go a long way in helping you deal with belligerent individuals.

Finally, communicate with your fellow employees on a consistent basis by using a communication vehicle such as a newsletter or website. These tools can provide answers sought by difficult people and encourage them to reach the conclusion that being difficult is harder and less productive than being a positive and satisfied person.

If you keep in mind that dealing with people is an essential part of doing business, you will have the first key to adjusting your attitude about difficult people. In some way, you must come to a realization that you genuinely like dealing with people. It is the people you deal with that make the decisions to use your product. You need to find a way to understand their needs. It will not serve anyone's needs to be at odds with "the hand that feeds you." You must find a way to foster and protect a positive attitude toward those you work with. This will, in turn, make you successful in dealing with all the people involved, including the difficult ones

CHAPTER 9 SUMMARY

This chapter discusses the importance of finding thought leaders in your organization, those sneezers or informal leaders who garner respect and can change an organization without a fight.

Use champions as:

- Recruiters for effort
- Points of contact for key areas
- Liasons between management and workers

Engage upper management:

- Integrate CPI with financial performance
- Conduct monthly management review meetings

- Do not forget to attach value to "soft" costs

Ensure annual goal commitments:

- Align with current measurement processes

Obtain executive team participation:

- Drives advocate and champion population

Deal with difficult people:

- Why are people so difficult?
  - Personality conflicts
  - Attitudes about change
  - Just having a bad day
- Finding out what is really bothering someone
  - Be a good listener
  - Get to the root cause
- Practical methods of handling difficult people
  - Be nonemotional
  - Make the person a part of the solution
  - Ensure that all agendas are clear, with clear outcomes noted
  - Emphasize communication

CHAPTER 9 AIDS—INFLUENCER WORKSHEET

Use the checklist on the next page as a way to help determine who the influencers are in your organization. Use these key areas as a guide or pull out the key areas of your improvement project and create headings. Then think about who can be of influence in that area and just how much influence you think they can garner, with 1 being very low and 10 being very high. This worksheet will help identify the heavy hitters in the organization and the people you will want to infect with your improvement virus so they can spread it around to everyone they meet.

Key Area	Person of Influence	Influence Rating	Comments
Design	Loren Averie, 15 yr Designer	10	Lots of respect, no process exposure
	Joshua Jacob	7	Excited about new processes; peers listen to her
Requirements	Kaedra Riann, SW Manager	9	Very exacting—has used requirements management tools
	Ashleigh Gardys, Test Manager	9	Open to requirements process—has had difficulty matching requirements specs to test
Configuration Management	Alyssa Michele	7	Process savvy, gets things done
Project Planning and Oversight	Devin Wayne	6	Has deep experience with project tools, but little experience in process

CHAPTER 10

World Climate—Checking for Vital Signs (Are there Any Signs of Life?)

The day I say, "This is good enough for me," is the day I begin to die. I want to know what I am. After that I want to know what I'm capable of becoming. And then I want to become it. I want to keep reaching higher all the time.
—Robert Silverberg, *Kingdoms of the Wall*

10.1 MEASURE, METRIC, INDICATOR

Every sporting event uses scorekeeping as a record of progress during the game or season. If the score were not kept during a game, how would a team or individual know whether it was winning or losing or how to strategize to win the game?

Most professional athletes also play close attention to their overall standings compared to their competitors. We guarantee that all professional golfers know exactly what score they need to make a cut. They understand what it takes to be successful and how to recognize when they have achieved a level of improvement.

From engineers to upper management, each individual will often speak of, with great clarity, the importance of measurement. At the same time, there seems to be a good deal of confusion about the terminology involved, specifically, *measure, metric,* and *indicator.* As Brice Ragland's article, "Measure, Metric, or Indicator: What's The Difference?", states, it is important to understand the differences between these terms [34].

Many people look to the Institute of Electrical and Electronics

Engineers, Inc. and the Software Engineering Institute definitions for guidance. What follows are some software terms and examples that will help to clarify the definitions.

First, look at the definitions of these terms:

Measure

> To ascertain or appraise by comparing to a standard. A standard or unit of measurement; the extent, dimensions, capacity, etc., of anything, especially as determined by a standard; an act or process of measuring; a result of measurement [35].

An example measure might be five centimeters. The centimeter is the standard, and five identifies how many multiples or fractions of the standard there are. Using the centimeter, someone measuring something in the United States is going to get the same measure as someone in Europe.

Let us relate this to software, such as lines of code. Currently, there really is no universal standard for lines of code. Someone measuring a program's lines of code in one office will probably not get the same count as someone measuring the same program in a different office. Therefore, it is imperative that each organization determine a single standard for what is meant by a line of code and ensure that everyone in the organization understands and uses that standard. A measure should be agreed on and understood as a standard. This standard should be defined specifically for the group that is using it. When doing this, be careful that the terminology used is also understood, so that the measure can truly be a standard, both for what it measures and the terms used to describe it.

Metric

> A quantitative measure of the degree to which a system, component, or process possesses a given attribute. A calculated or composite indicator based upon two or more measures. A quantified measure of the degree to which a system, component, or process possesses a given attribute [36].

An example of a metric would be that there were only two user-discovered errors in the first 18 months of operation. This provides more meaningful information than a statement that the delivered system is of top quality.

Indicator

> A device or variable that can be set to a prescribed state
> based on the results of a process or the occurrence of a
> specified condition, i.e., a flag or semaphore. A metric that
> provides insight into software development processes and
> software process improvement activities concerning goal
> attainment [36].

As the definition notes, a flag is one example of an indicator. An
indicator is something that draws a person's attention to a particular situation. Another example of an indicator is the activation of
a smoke detector in your home; it is set to a prescribed state and
sounds an alarm if the number of smoke particles in the air ex-
ceeds the specified conditions for the state for which the detector
is set. In software terms, an indicator may be a substantial increase in the number of defects found in the most recent release of
code.

The objective is not to add more definitions or confusion, but to
give an example to help understand the differences between these
terms. A few charts can help clarify the differences. Let us start
with a common scenario that involves a sick patient.

An individual is brought into a hospital emergency room. He is
unconscious and has a temperature of 99.1 degrees Fahrenheit, as
shown in Figure 10.1. Other vital signs appear normal. What does
the *measure* of 99.1 degrees Fahrenheit tell you? Very little. You
may realize that it is above normal body temperature, but you do
not know if the temperature is going up, down, or remaining constant. So is this individual getting better or getting worse?

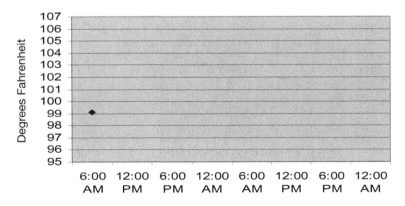

Figure 10.1. Body temperature (measure).

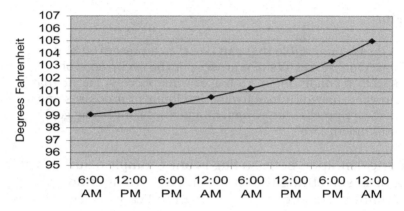

Figure 10.2. Body temperature (metric).

Now, after many hours of regularly checking the patient's vital statistics, you are able to see a trend in the temperature readings, as shown in Figure 10.2.

This trend analysis gives the doctors a lot more to work with, even though the patient is still unconscious. What does the chart in Figure 10.2 show us? The temperature continues to climb even more rapidly as the second day progresses. The doctors start to worry, but other vital statistics show no problem.

Suddenly, the patient awakes and provides more information about his condition. His is from the planet Mars, and his normal body temperature is 105.6 degrees Fahrenheit, as shown in Figure 10.3. He was recovering from hypothermia.

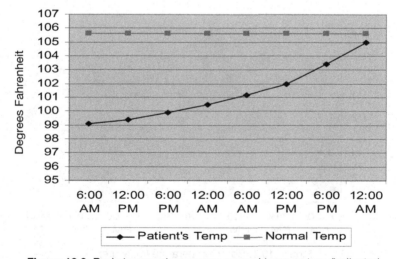

Figure 10.3. Body temperature versus normal temperature (indicator).

This scenario helps to illustrate the difference between measures, metrics, and indicators. Figure 10.1 shows a measure. Without a trend to follow or an expected value to compare against, a *measure* gives little or no information. It especially does not provide enough information to make meaningful decisions. Figure 10.2 shows a *metric.* A metric is a comparison of two or more measures, in this case body temperature over time, or defects per thousand source lines of code. Figure 10.3 illustrates an indicator. An *indicator* generally compares a metric with a baseline or expected result. This allows the decision makers to make a quick comparison that can provide a perspective as to the health of a particular aspect of the project. In this case, being able to compare the change in body temperature to the normal body temperature makes a big difference in determining what kind of treatment, if any, the patient may require.

This example is obviously fictitious, but it does illustrate the point that a little bit of information can be dangerous. This does not mean that no information is better; it means that you need the right amount of information of the right kind to make the best decisions. Do you wait until you have all the information before you make a decision? No. But recognize that without enough of the right information, there is a risk involved in making that decision.

The example also illustrates that your frame of reference is not always the right one. You must be willing to look at situations with an objective view. If you cannot see a situation from more than one angle, you may need to request consultation from someone with a different perspective.

10.2 INDICATORS OF SUCCESS

To be successful, a company must measure and report its performance on a routine basis. When designed and implemented effectively, performance measurement does the following:

- Supports the organization's strategic plan by providing management with tangible indicators and goals relevant to daily activities.
- Provides executives with sufficient and timely information regarding the effectiveness of operations before significant financial impacts are experienced.
- Creates a work environment that supports and rewards cooperation among key functional areas to obtain desired results.
- Drives change by focusing resources and shaping behaviors toward specific, tangible, results.

- Establishes a mechanism for assigning and enforcing accountability as well as recognizing and rewarding outstanding performance.

If your effort revolves around the Software Engineering Institute models, measure your progress according to that scale. Even if your management feels like a group should never have its capability published, doing so will likely have a positive effect on your organization. Count the compliers and paint them vibrant red, yellow, and green colors. Set up assessment polls on your internal website and post green/yellow/red status charts, as shown in Figure 10.4 (of course, the print version of this book is black and white, so yellow is the lightest tone, green is the middle tone, and red is the darkest tone). Everyone across the organization can view these results. Before you know it, projects will be vying for the top spot. You can measure compliance by percentage and tally the results. Many organizations share this information only with the team, but you may find that your culture may thrive on the competition. First, they will be excited to get all green. Then they will compete to have the highest percentage for compliance. It adds a lot of fun to the metrics experience to announce weekly status.

There will be other benefits of posting results; for instance, one of them may be unforeseen. First, as expected, teams that are not doing well in the metric race will seek out teams who are doing well and get mentoring from them. They will borrow people from the other teams to come in and help them out in a particular area. They will also discuss best practices during weekly meetings and share what they have learned, which will further accelerate your ability to improve your processes.

An unforeseen benefit will be that teams start tracking each other as an advocate would. They will pressure each other to perform better and will open up discussions if they feel that a team is

	RM	PP	PTO	SSM	QA	CM	
Project A							83%
Project B							70%
Project C							76%
Project D							68%

Figure 10.4. Stoplight chart.

under- or overperforming in relation to their scores. This builds in a good form of peer pressure that further strengthens your organization's ability to grow and improve your process.

10.3   RETURN-ON-INVESTMENT COMPUTATIONS

Raytheon and Hughes Aircraft have compiled and published the returns they have received on their process improvement investments. In Ray Dion's article titled "Process Improvement and the Corporate Balance Sheet," Raytheon showed a two-fold increase in its productivity and a 7.7-to-1 return on its improvement expenditures for a savings of $4.48 million on a $0.58 million investment [39]. Over a four and one-half-year period, the company eliminated $15.8 million in rework costs.

Watts Humphrey, Terry Synder, and Ronald Willis published an article titled "Software Process Improvement at Hughes Aircraft," which stated that Hughes Aircraft has computed a 5-to-1 return on its process improvement initiatives, based on changes in its cost performance index, which is the budgeted cost of work performed divided by the actual cost of work performed [38]. The company has experienced an annual savings of approximately $2 million due to its process improvement expenditures.

10.4   SPILLOVER BENEFITS OF PROCESS IMPROVEMENT

The benefit most frequently noted concerns attitudinal changes. The morale and confidence of the process users improves significantly, and the process users experienced increased attention and respect from organizations external to theirs. Users also noted less overtime, less employee turnover, improved competitive advantage, and increased cooperation between functional groups as benefits that result from process improvement initiatives.

CHAPTER 10 SUMMARY

- Measure, metric, indicator
  - Measure: Appraise by comparison to a standard. Example: lines of code.
  - Metric: A quantitative measure of a given attribute. Example: Number of user-discovered errors.
  - Indicator: A variable that can be set as a flag to signify a

specified condition. Example: A substantial increase in the the number of defects found in the most recent release of code.

- Indicators of success; performance measurements
  - ○ Displays status against given goals
  - ○ Supports strategic plan
  - ○ Provides visibilty
  - ○ Rewards cooperation
  - ○ Helps shape behaviors
  - ○ Enforces accountability
- ROI computations
  - ○ Raytheon reports productivity increases of 7.7 to 1 on investment
  - ○ Hughes Aircraft reports a 5-to-1 return
- Spillover benefits of process improvement
  - ○ Better attitude
  - ○ Increased morale
  - ○ Higher job satisfaction

CHAPTER 10 AIDS—SAMPLE MEASUREMENT PLANNING WORKSHEET

The following example is from an actual subteam that documented how to extract metrics for project tracking effectiveness. This outline gives you a sample of the total process of building reasonable metrics for the areas you are trying to track. In this example, you can see how one process area was analyzed and the metrics that were tracked as a result of this exercise. It is useful to document the reasoning behind a metric, so that an organization can review their metrics and assumptions periodically. The arrows indicate the "hot issues" for the organization.

SUMMARY OF PROCEDURE DOCUMENT

Project Tracking and Oversight Policy

- **Purpose**
  To achieve adequate visibility into actual progress.

- **Requirements**

  →   1. **Actual performance and results are tracked** against project plans and analyzed.
  2. Corrective actions are taken and managed to closure.
  →   3. **Changes to software commitments are agreed upon.**

- **Verification**

  →   1. **Review status** of the project both on a **periodic** and event-driven basis.
  2. **Summarize the status** on a **monthly basis.**
  →   3. Review the activities and work products.

RECOMMENDED METRICS

(To reinforce the project tracking and oversight policy)

     1. Need to establish (develop) a project plan
         Parts of the plan are:
            Software requirements
            Software configuration management
            Software quality assurance
→   *Schedule (project tracking and oversight). This is our targeted metric. We will measure schedule.*
     2. Define the product deliverable categories (part of the project plan)
            Alpha version
            Beta versions
            Other (Early adopter, inner circle, etc.)
            First customer ship (FCS)
            Metric one
               Record any change in definitions:
               New definition:
               Date changed:
     3. Establish Schedule

Metrics two/three

Define at analysis phase:

   Analysis start/complete—original plan; current plan; actual

   Design complete—original plan; current plan; actual

   Implementation complete—original plan; current plan; actual

   Alpha original plan—current plan; actual

   Beta versions—original plan; current plan; actual

   First Customer Ship—original plan; current plan; actual

Phase schedule, comparison of the plan to actual

A number is generated that is the performance-to-phase schedule (for each phase) = original plan date – current plan date (in months)

   Example: Analysis complete slipped Jan to March = +2

   Design complete slipped April to July = +3

   If dates are undefined at the Analysis Phase project is put in the category of "not to plan."

Deliverable schedule—comparison of the plan to actual

A number is generated that is the performance-to-product schedule = original plan date – current plan date

   Example: First customer ship—early July to June = –1

Program:_____

Define the product deliverable categories

Original plan date:_____

Deliverables	Yes	No
Alpha version		
Beta versions		
Other (early adopter, inner circle, etc.)		

First customer ship	Date:	

Revised plan date:_____

(Record Changes—how many changes occurred during this project?)

Total number of changes equals _____

Project/Program Start Date:

Benchmarks	Orig. Date	Current Date										Actual Date Achieved	Notes
		t1	t2	t3	t4	t5	t6	t7	t8	t9			
Analysis Start													
Analysis Complete													
Design Complete													
Implement Complete													
Trial Complete													
Alpha Delivered													
Beta 1 Delivered													
Beta 2 Delivered													
Other Delivered													
FCS													

Who will be recording the data?

Department managers, program managers, or designees record the baseline data at the analysis gate. Data is subsequently updated at least monthly or more often, as driven by events.

How will the data be used?

1. The data will be reported on the operations report under the heading "Performance to Phase Schedule." It will be used as an input to judge the stability of the program, as an input into the lessons-learned analysis, and as an input to the development-time database.
2. The data will be reported on the operations report under the heading "Performance-to-Product Schedule." It will be used as an input to judge the stability of the program, as an input into the lessons-learned analysis, and as an input to the development-time database.

CHAPTER 11

World Health— Evaluating Progress

Trust, but verify.
—Ronald Reagan

11.1 EVALUATING YOUR PROGRESS

Checking the progress of your efforts will necessitate tracking the progress of not only your champions and advocates, but also the process itself. Since the actions of your change leaders will affect how the teams perform, it will be important to take a close look at how they are contributing to team performance and affecting overall process-improvement progress. Once you acknowledge that your champions and advocates can greatly affect your team's performance, you will be able to measure their progress in meaningful ways. In turn, this evaluation will further support your push to improve continually.

This is an example of an area in which your Director or Vice President of Quality can be of assistance. Your Quality representative can help form the wording that should go into evaluations and can also prepare the evaluation of the process final report. This person can also validate any needed corrective actions and verify that there is closure to any issues and training for any gaps in process knowledge.

Collaborative Process Improvement. By C. L. Yeakley and J. D. Fiebrich
Copyright © 2007 IEEE Computer Society.

11.2 EVALUATION OF CHAMPIONS AND ADVOCATES

Just as each person has a unique fingerprint, there is uniqueness in each one of your personnel. Evaluations are essential to getting every person's attention. Because of your efforts to keep management informed and because you have specific and measurable goals to obtain, it is relatively easy to get all supervisors to add process-improvement activities to yearly goals. People who cannot show how they contributed to the process improvements stated in the annual goals should not be rewarded as highly as people who make notable contributions. Furthermore, it will be obvious that teams that engage in the process improvement activities perform better.

Comparisons of project team performance will reveal that teams that have an active advocate perform better by delivering better products closer to the promised time. Each advocate must be engaged in giving constant feedback to the team by reporting on compliance to specific activities. Teams can benchmark themselves against others as well as compare their own performance to that in the months prior to the improvements.

It is also very important to note that, as in all good leadership philosophies, you want to leverage strengths and minimize weaknesses. Be careful in all evaluations by looking for ways to build up rather than to tear down. In this age of the controversial "rank and yank" or "vitality curve" companies, it is important to understand the inner workings of your teams and to encourage teamwork. It is very hard to encourage teamwork in an atmosphere of ranking and rating of employees that is at best only effective for a round or two of evaluations. Once the initial rounds of these sorts of evaluation systems have been completed and the bottom 10% have been encouraged to move on, there is a tendency to deemphasize teamwork, thereby causing distrust and major politicking. You must determine ways to encourage team performance by making it important to everyone's evaluation.

11.3 EVALUATION OF THE PROCESS

After you have considered the role of champions and Advocates in your process, you will need to have a method to determine the extent of your organization's improvements. The key to improving an organization's performance is to understand its current situation by capturing and analyzing its process measurements. Once an organization understands the strengths and weaknesses of its current process, it is in a position to effectively apply resources toward improving its performance. Without such an understanding, attempts to improve may be ineffective or, even worse, counterproductive.

11.4 MONITOR THE IMPROVED PROCESS

Once you have taken action to make a significant change to one of the processes, the next task is to gather objective evidence to validate that expected improvements are indeed taking place. One major problem you must overcome is the excessive time that is frequently needed before reliable feedback can occur. For example, it is not possible to compute the percentage effectiveness of the processes until a new release has been in operation for several years. This definitive measurement is useful for studying long-term trends and for confirming that a process is under statistical control, but by the time it is available, the data is too stale to be of practical use in driving a process-improvement program. Once a process is under statistical control, it is possible to map its trends and establish confidence boundaries that you can use in predicting future performance. Then, as the process is executed, the interim results can be compared to the confidence boundaries and conclusions drawn about the effectiveness of the process in comparison to the past.

11.5 CONDUCTING AN EVALUATION

You can divide a process evaluation into the preparation phase, performance (execution) phase, reporting phase, and the corrective-action follow-up phase [39].

11.5.1 Preparation Phase

The preparation phase starts when the decision is made to conduct an evaluation. It includes activities such as team selection, on-site information gathering and review, and communicating the evaluation plan and audit agenda to the party being evaluated. Between the time an evaluation is assigned until the actual start of the evaluation, you should follow the following steps:

- Define the purpose of the evaluation
- Define the scope of the evaluation
- Determine the evaluation-team resources to be used

- Identify the authority for the evaluation
- Identify the performance standards to be used
- Develop a technical understanding of the process to be evaluated
- Contact those to be evaluated
- Perform an initial evaluation of lower-tier documents to higher-level requirements
- Develop written checklists of the data needed

The outputs of the preparation phase are an evaluation notification letter, an evaluation plan, an evaluation agenda, an evaluation checklist, logistical arrangements completed, initial evaluation of the control methods, and a plan for the collecting of facts.

Instructional designer Kim Hansen insists that you must be careful to keep this as simple as possible. Also pay particular attention to the language used. In the performance improvement methodology called Appreciative Inquiry [42], evaluations are used by groups to explore what is working well in an organization so that they can discover further positive improvements. Evaluations then would be the equivalent of Appreciative Inquiry stage 1, Inquire, "Appreciating the best of what is." People tell stories with positive reflection so that they appreciate what they have accomplished. The point is that evaluations should be worded in positive terms rather than negative terms. This is a "qualifier" evaluation method rather than a "disqualifier" method. This qualifier method looks for positive attributes of people and processes rather than looking for things that could be considered weaknesses. Be sure to word your evaluation checklists in a positive light, for example, "What were your greatest successes?" rather than "What was the largest error made on the project?"

11.5.2 Performance Phase

The performance phase begins with the on-site opening meeting, and includes the gathering of information and analysis of that information. Normally, this is accomplished by conducting interviews, monitoring activities, and examining items and records. This phase is completed when the exit or closing meeting is held.

The performance phase of an evaluation is often called the field work. It is the data gathering and analysis portion of the evaluation, which includes the opening meeting, understanding the software development process and controls, communicating with evaluation-team members, interviewing and briefing individuals to be evaluated, and the exit meeting or draft report.

Findings and observations are to be presented in an impersonal and factual manner. The evaluation-team meetings are used to reach consensus on findings based on the evidence. A finding is an evaluation conclusion that identifies a condition having a significant adverse effect on the quality of the activities under review. Each finding must be a clear, concise statement of a generic problem.

11.5.3 Reporting Phase

The reporting phase covers the translation of the evaluation team's findings and conclusions into a formal evaluation report. It is generally accepted that the evaluation report should be accurate, concise, clear, and timely (i.e., mailed within one week). The evaluation report and pertinent documents are retained for a specified time period. The report should reflect both the tone and content of the evaluation and be consistent with the results presented in the exit meeting. This phase covers activities such as prioritizing the evaluation results; preparing the evaluation report; listing findings and observations, including corrective action requests; expressing a judgment of the extent of the evaluation's compliance; approving and distributing the report; and retaining evaluation records.

Evaluation records may be classified as either long- or short-term records, depending on their use and the length of time they are kept. Although there is much variance in the industry, a good length of time for keeping long-term evaluation records is five years. However, the evaluation plan should provide the required retention period.

11.5.4 Corrective-Action Follow-up Phase

The group being evaluated is responsible for providing the lead evaluator with a corrective action plan within a specified time frame. The plan will outline the actions to be taken by the evaluated party to rectify the noncompliance, the individual assigned, and the estimated date of completion. Some form of formal closure criteria should be stipulated by the evaluated party, with which the evaluator can verify successful implementation during the follow-up evaluation. The corrective action is considered by the evaluator for closure once the corrective action has been implemented and verified.

The corrective action undertaken must demonstrate that it was implemented in a timely fashion, is effective, and is designed to prevent further reoccurrences. The steps leading to a formal closure of the corrective-action request might necessitate another site visit to verify implementation and effectiveness of the corrective action.

This phase covers the following activities:

Proposed corrective-action plan (evaluated party)

- Investigates possible causes, strives to eliminate root cause
- Identifies preventive action or improvement
- Establishes closure criteria
- Develops a corrective-action plan for each finding
- Submits corrective-action plan for evaluator review and approval
- Evaluates and approves evaluated party's proposal plan

Implementation of corrective-action plan (evaluated party)

- Implements improvements based on corrective-action plan
- Releases new or revised process and/or procedure documents
- Provides training on improvements to affected personnel
- Demonstrates implementation of improvements on subsequent projects
- Advises evaluator when improvements are completed and ready for verification of completion and effectiveness

Follow-up (evaluator)

- Tracks the planned completion dates for corrective action and follows up with the evaluated team to determine if changes have been completed
- Schedules follow-up evaluations after change is implemented and permanent, and the new process is being used by the organization or project
- Determines whether corrective action was implemented and, if it is effective, generates a brief report that states that the evaluation is either still open or is now satisfied and closed
- Negotiates a renewed effort with the evaluated team when the results of corrective action appear to be ineffective
- Escalates the situation to the client's upper management when plans are not available (i.e., within 30 days) or estimated completion dates are not met

Criteria for closure (evaluator)

- Considers closing the corrective-action request after the corrective-action has been implemented and verified
- Makes a decision as to whether the corrective action taken fixed the symptom or, indeed, corrected the root cause of the deficiency
- Follows the closure process defined in the organization's corrective-action-request procedure
- Schedules corrective action for examination at the next regularly scheduled evaluation after completion of the corrective-action request
- Retains the closure documentation as part of the corrective-action request history
- Closes the evaluation by letter or memo after all findings have been closed

11.6 BAD RESULTS

Author Chris Argyris discloses in his book *Flawed Advice and the Management Trap* [41] that bad results from a process assessment can quickly put personnel on the defensive. Once bad results put people on the defensive, they usually default to behaviors learned in early childhood to avoid embarrassment. People around them notice the defensive reaction and no longer discuss whatever created the defensiveness.

Those on the defensive shape discussions about results so that they look good, or at least do not look bad, by making excuses, laying blame, or making attributions, all of which are kept general and untestable. Once no one mentions certain embarrassing facts, the situation becomes self-sealing [41].

To avoid self-sealing situations during process assessments,

- Keep discussions about processes factual and testable. Accompany any bad news from process assessments, with one or two specific proposals for process-improvement teams for corrective action.
- Turn the deployment of all new process assessments into opportunities to provide training and coaching in understanding variation [42], process-improvement tools, and teaming, as discussed in Donald Wheeler's book, *Understanding Variation.*
- Check your assessments regularly to ensure that each mea-

sure is still aligned with the organizational mission, strategy, or top priorities.

CHAPTER 11 SUMMARY

Addressing evaluations is a potentially difficult exercise. In this chapter, we have looked at both the role of evaluating champions and advocates and the means of tracking improvements. The key points are:

- Evaluate champions and advocates
- Promote and understand their role in team performance
- Leverage their strengths and minimize weaknesses
- Evaluate the process
- Monitor—gather objective evidence and track trends,
- Conduct an evaluation in phases
- Preparation—selecting the team, gathering information, review, and communication. Assemble the results in a plan checklist and consider "appreciative inquiry" methods.
- Performance—includes gathering information and analysis. Often called field work.
- Reporting—translation of findings and conclusion into an evaluation report.
- Corrective action (if necessary), should have a plan for correction within a specified time frame. Includes preventive actions. Implementation of the plan could include training, prevention, and follow-up.
- Bad results—avoid putting personnel on the defensive. To do this:
  - Keep discussions factual
  - Turn new process assessments into opportunities for training and coaching
  - Ensure alignment with corporate strategies

CHAPTER 11 AIDS—ASSESSMENT CHECKLIST

Use this checklist to help you to prepare and conduct an internal assessment of your processes.

- Have the team members been briefed on the process scope (what is to be reviewed and what is not to be reviewed) and areas to be assessed?

- Do you have people assigned to help with the assessment? (Consider using champions as assessors.)
- What is the assessment schedule?
- Does the team understand what documentation/materials your assessment team would like to review?
- Are your assessors trained to perform a process review?
- Do you have a format for the final report?
- Are assessment checklists available?
- Is the team that is being assessed prepared to participate in the assessment process?
- Is there a single point of contact that is prepared to track the resolution of any issues found during the assessment?
- Are attitudes about the assessment positive?
- Have you determined a way to share best practices discovered as a result of the assessment?

CHAPTER 12

World News—Rewarding and Recognizing Work

You can buy a person's hands, but you can't buy his heart.
His heart is where his enthusiasm, his loyalty is.
—Stephen Covey

12.1 REWARD PROGRAM

Every company needs a strategic reward system for employees that address these four areas: compensation, benefits, recognition, and appreciation. The problem with reward systems in many businesses today is twofold: They're missing one or more of these elements (usually recognition and/or appreciation), and the elements that *are* addressed are not properly aligned with the company's other corporate strategies. Frankly, some of the rewards handed out seem, for lack of a better word, heartless. How many times have you seen it happen that people do not even post their "award" on their cube or office wall? Have you ever wondered why people do not care? In order for a reward to mean something to the employee, it needs to be something that touches the heart. It needs to be something that gives that "warm and fuzzy" feeling of a homemade pie. It is the difference between getting a store-bought pie and being the recipient of a hand-made, fresh-out-of-the oven, grandma's pie.

A winning system should recognize and reward two types of employee activity—performance and behavior. Performance is the easiest to address because of the direct link between the initial goals you set for your employees and the final outcomes that result. For example, you could implement an incentive plan or recognize your designer for reaching periodic goals.

12.1.1 Rewarding Behaviors

Rewarding specific behaviors that make a difference to your company is more challenging than rewarding performance, but you can overcome that obstacle by asking, "What am I compensating my employees for?" and "What are the behaviors I want to reward?" For example, are you compensating employees for coming in as early as possible and staying late, or for coming up with new ideas on how to complete their work more efficiently and effectively? In other words, are you compensating someone for innovation or for the amount of time they are sitting at a desk? There is obviously a big difference between the two. Slaving away at one's desk is not the thing that dreams are made of. People get excited by innovation and creativity. A commonly made mistake is when an employee gets an award that is presented in front of a group of fellow employees and includes the words "for putting in many hours of overtime." The reward can be seen as being based on sheer magnitude of hours, or so it would seem by the glowing statements of accomplishment that include hours on the job.

The first step in determining who should be rewarded, of course, is to identify the behaviors that are important to your company. Those activities might include enhancing customer relationships, fine-tuning critical processes, or helping employees expand their managerial skills. Remember that anyone can sit at a desk for hours on end and look haggard. It is amusing to listen to employees' comments about rewards based on "hard work." *No one* will admit that they work a 40-hour week. It is as if they are all in a competition about how overworked they are. In reality, particularly in the software development world, a tired, overworked employee is the one that makes the most mistakes and contributes to poor quality of the end product!

When business owners think of reward systems, they typically put compensation at the top of the list. There is nothing wrong with that, since few people are willing or able to work for free, although many who are independently wealthy will continue to work because they find their work fun and challenging. The right strategy should also include an incentive-compensation plan that is directly linked to the goals of your company for that period. You might want to include some type of longer-term rewards for key individuals in your organization. Historically, this has often included some form of equity ownership.

Benefits are another type of reward in a strategic reward system, and your employees are definitely going to notice the types of benefits you provide. Companies that do not match or exceed the benefit levels of their competitors will have difficulty attracting and retaining top workers. This is one reason an increasing number of businesses are turning to professional employer organizations to gain access to a broader array of company benefits.

12.2 RECOGNITION AND APPRECIATION

Having said all we have about rewards, you still cannot diminish the importance of recognition and appreciation as integral components of a winning strategic reward system. These two elements rarely receive the attention they deserve from management, which is amazing because they are the low-cost/high-return ingredients. Employees like to know whether they are doing good, bad, or average, so it is important that you tell them. Sometimes it seems that if a solution is easy, it is invalid. It is easy to overlook the simple answer, but do not do it! People appreciate recognition!

Recognition means acknowledging someone before his or her peers for specific accomplishments achieved, actions taken, or attitudes exemplified through their behavior. Appreciation, on the other hand, centers on expressing gratitude to someone for his or her actions. Showing appreciation to your employees by acknowledging excellent performance and the kind of behavior you want to encourage is best done through simple expressions and statements. For example, you might send a personal note or stop by the employee's desk to convey your appreciation. Another approach is to combine recognition and appreciation in the form of a public statement of thanks in front of the employee's coworkers or team, citing specific examples of what they have done that has positively impacted the organization, but be careful not to attach that "hard work" moniker to what they have done. If you want people to work smart, be innovative, and feel that change is important, you have to use words that support those concepts.

12.3 REWARDING ADVOCATES AND CHAMPIONS

Do not ever be hesitant to reward your advocates' and champions' efforts. Most people have little idea what makes them happy, motivates them, or gives them lasting satisfaction. Use both intrinsic and extrinsic rewards and recognition throughout the year. Intrinsic rewards and recognition include public recognition, trophies, plaques, certificates, special parking spaces, pictures on bulletin boards, and names on a list. Extrinsic rewards and recognition include cash in excess of $1000, vacation time, bonuses, and major gifts. When initially planning the types of rewards for the advocates and champions, 80% of the advocates will say they prefer extrinsic rewards: "Money is what pays my bills!" However, six months later, being presented with a certificate during a staff meeting will be the desire of the majority of the advocates and champions. Team rewards, which can be intrinsic or extrinsic items given to the team, should be the same for all members of the team.

The ultimate reason that rewards and recognition are given is to provide positive reinforcement for correct behavior, with the expectation that the correct behavior will be repeated in the future. Rewards and recognition are best received when they are personal to the individual receiving them. Of course, do not forget to look first at this group of individuals when good positions are available. One of the best ways to get an organization's attention is to promote someone from your advocate or champion group. Nothing implies success like a promotion.

Twelve months into the advocate program, when polling the advocates to determine whether they prefer an extrinsic reward such as cash, stock, or stock options, or an intrinsic reward such as having a director discuss an individual's accomplishments in an operations meeting, 100% will prefer the latter.

As a matter of fact, at least once a week, we had individuals approach us and state, "I do not care about a cash bonus; I want to know what I have to do to have the director discuss my accomplishments for ten minutes at the next operations meeting." And, of course, this reward costs the business nothing.

Probably one of the best rewards is a "thank you," when it is sincerely meant. Employees who are aware that their efforts are appreciated are often willing to do more than if they were to receive a large financial reward.

Rewards and recognition are best received when they are personal to the individual receiving them. This goes back to the heart of recognition. Are you just going through the motions of rewarding someone because it is expected, or do you really *care* that this person made a significant difference in the life of your company? Reward distribution is more difficult to achieve as the team size increases or as the number of awards increases. If the award is unique, it has greater value than the same recognition a second or third time. If the company gives out a certificate of achievement to one-third of the employees each month, the employees will very quickly think of the certificates as worthless and a waste of time.

Recognition by management of a team's successful completion of the team objectives can be very positive, and can encourage other teams to strive for excellence. If the team's efforts are viewed by other teams as being less significant contributions than were made by teams not receiving recognition, this can be counterproductive. Team and individual rewards and recognition are important to both the organization and to the individuals, but are difficult to implement correctly. Although a reward may motivate one group, failure to receive a similar reward may discourage another group. The same reward, given under different conditions and at different times, may have totally different results or acceptance.

Examples of small rewards are:

- The team that finds the most errors in the product receives a gift card for a local-area cinema.
- Dinner certificates are given the advocates who are voted Advocate of the Month.
- An ice cream social afternoon is held for the improvement team.
- The best operating team can select their next project.
- Employee of the month parking spot near the entrance.
- The most improved team is given an afternoon at the movies, bowling alley, or skating rink.

CHAPTER 12 SUMMARY

- Rewarding Behaviors
  - What behaviors do you want to reward?
  - Incentive compensation plans
  - Longer-term rewards
  - Be competitive with other organizations
- Recognition and Appreciation
  - Simple is still valid
  - Recognition involves acknowledgment in front of peers

○ Appreciation can be accomplished through simple expressions and statements
- Rewarding Advocates and Champions
  ○ Intrinsic awards—discussion of accomplishments in a public forum, positive statements, "thank-yous"
  ○ Extrainsic awards—cash, vacation time, bonuses, major gifts
  ○ All should be personal to the individual
  ○ Team rewards can be positive and influence other teams to perform

CHAPTER 12 AIDS—SAMPLE CERTIFICATES

Certificate of Appreciation

This Certificate is awarded to:

_____

For STAR Performance in Collaborative Process Improvement
Activities

We recognize STAR performance:
Special efforts
That result in
Astounding
Results

Signed _____ (CEO), Date _____

Signed _____(CPI sponsor), Date _____

Employee of the Month

[Company Name] recognizes:

_____ as Collaborative Process Improvement employee of the month for working smart and building bridges for process improvements.

The bearer of this certificate is entitle to the Employee of the Month parking spot for the month of _____, 20XX.

Signed _____ (CEO), Date _____

Signed _____(CPI sponsor), Date _____

Certificate of Appreciation

[Company Name] recognizes:

_____ In appreciation for Collaborative Process Improvement efforts in support of achieving Company Strategies.

The bearer of this certificate is entitled to lunch with the CEO (to be scheduled).

Signed _____ (CEO), Date _____

Signed _____(CPI sponsor), Date _____

CHAPTER 13

The Modern World— Building Meaningful Quality Pictures

You must take the Customer's skin temperature every day.
—Konosuke Matsushita, Matsushita Corporation

13.1 TEAMWORK

In order to build a CPI-focused organization, you will want to engrain other supporting elements so that your CPI efforts will stick. Teamwork is an essential element in CPI. It is hard to collaborate without people, and people who are fun to work with make the work seem light. A workplace that is fun can relieve stress, gel teams, and be more productive and creative. Hopefully, the self-forming teams that will be launched as a result of a CPI effort will be the sort of interesting and fun teams that people will want to engage in. You can help foster this teamwork by laying the groundwork for team building and understanding individuals in teams. This can be done as part of the start-up activities of the overall CPI project or at any time a new team is formed.

Part of what makes a team fun and easy to work with is an understanding of each other and trust. Where there is no understanding, there is generally a lack of trust. For this reason, engage in some of those silly team-building exercises that people initially find boring. A good exercise to do that is quick to use to demonstrate different personality types is to ask about requirements. Ask, "How do you feel about requirements? Are the requirements of a product met if everything that was contracted to was achieved?" Listen to the different answers. In a large group, you will likely get a variety of answers like "Yes! That is what we contracted to," to "It depends," to "No, if the customer does not like the final product, we have missed our requirements, no matter what our specifications were!"

Another exercise to use (and is particularly appropriate in jump starting nontechnical projects) is the "Instant Vacation" game. It goes like this: "Congratulations! You have just won $15,000 to take a vacation. You will have two weeks time off. Tell me about your trip! What will you do? Where will you go? Will you pack ahead of time? Tell me about how you would plan for your trip."

The point of this exercise is to demonstrate how different personality types will approach this problem. The detail-oriented person will want to have plenty of time to plan. They will have lists. They will have lists for packing, lists for what to do on vacation minus 90 days, vacation minus 60 days, and so on. They will not be comfortable with the thought of taking a vacation without a lot of planning to maximize their experience. They will also plan what they will do every day. These are the same people who are likely to say that the requirements are whatever is documented and nothing more.

Then you have the other end of the spectrum. Some people will take off right away and get in their car, turn the key, and go—without even a map. They will just go where they feel like going and figure out what to do along the way. Pay special attention to the faces of the other team members and to any questions that might arise. This will tell you a lot about your group. For example, there will be people who will ask follow up questions (these are generally the detail-oriented people) like "You mean you will just jump in the car? Without even a map?" to which the spontaneous person answers, "I will drive until I feel like it is time to stop." Just listening to the discourse when you open up an exercise like this will be enlightening.

The key lesson that teams can take away from this sort of exercise is that it takes all types to build a successful product. Just as the person who is very detailed and has almost endless lists is not wrong, neither is the person who just jumps in the car, turns the key, and goes. If your entire team is composed of people who are one type, who *must* have a list, then you need to find people of different personality types in order to build a fun and effective team. Just as the body needs more than just hands—it needs eyes, ears, feet, a brain, and so on—your teams must have people of different types.

Doing exercises like the above will serve two purposes:

1. It will point out that people *are* different and open up discussion on how best to work together.
2. It will be a team-building experience that you can share.

To have an effective team, everyone will need to learn how to minimize people's weaknesses and use people's strengths. In the example above, which person would you prefer to have tracking tasks or monitoring contracts? Which person would you prefer to have acting on a sudden change in plans or requirements? You will be teaching your team to work well together, leverage their strengths, and build in the capability to collaborate effectively.

Part of the continuous nature of process improvement can include the improvement of the teams themselves. Teams should be coached to take time out to evaluate how they are operating. Doing a postmortem is not quite enough to keep a team operating well. Several checkpoints can be included over whatever time period the team is operating together.

13.2 QUALITY WITH HEART

How does an organization build a picture of what it means to produce a quality product? The word "quality" is subjective, and measuring subjective factors is rife with pitfalls. Since quality is in the "eye of the customer," just as beauty is in the "eye of the beholder," how does a good quality manager or representative get to the heart of the customer and develop a set of criteria that make sense in their customer's world?

In order to make Quality real with your employees, so that they will take your CPI methods to heart and hold the customer in their daily visions at work, you can help build a human definition of what your customer looks like. When you know what the customer looks like, then the definition of a quality product will become clearer to each employee, and they can improve overall processes to create a product that will delight your customers. If you get too involved in just improving the process without considering the "face of the customer," you could lose valuable market share. If you can, drive the idea home by having customer seminars at which you expose your customers to your employees in a non-threatening way. You can even post photos of your product(s) at home with the customer to consistently drive the idea that customers are the ultimate users of anything that you are building and help everyone understand the problems you are trying to solve. If you can understand their problem and/or alleviate their pain, then you will capture the heart of the customer.

Many organizations struggle to define quality in terms that are meaningful translations of what customers are asking for. The challenge is to build a set of quality attributes that are applicable to your particular organization or, perhaps more aptly, your particular customer's problem space.

This section will outline some ways that you can get to the heart of your customer because, after all, if you capture your customer's heart, you have also captured your market. That is the "right stuff" that will give you a competitive edge in a tough marketplace.

13.2.1 Ask these Questions

How do you fix the problem of potentially being out of touch with the customer? Basically, you will want to get an answer to "what is Quality?" for your customer in terms that can be understood well enough that a product can be developed that has the potential of delighting your customer base. Textbooks are full of general Quality measures. Roger Pressman, author of one of the most widely used textbooks in all of software engineering *Software Engineering, A Practitioner's Approach,* [45] also maintains a website that has more links than a mere mortal can peruse in a year or two.

It is always good to go to the experts to get answers to your questions, but it is also valuable to go to your customers to get answers to their quality questions. So, know your "stuff" and be prepared to put on your customers' shoes (or better yet, get behind their eyes) so that you can see what they see.

IEEE Standard 1465-1998 divides quality requirements into three areas:

1. Product—Contents, indentifications, indications, and statements on functionality, reliability, usability, efficiency, maintainability, and portability

2. Documentation—Completeness, correctness, consistency, understandability, ease of overview

3. Programs and data—functionality, reliability, usability, efficiency, maintainability, and portability

You are likely to get a very tepid answer if you just ask your customers directly what they consider to be a quality product. "So, Mr. Customer, what does quality mean to you?" After the sound of air being sucked in quickly, your customer is likely to say, "I want a good product, on time, that does what I want it to do!"

That is hard to measure! Therefore, you need to get to the root of what the customer wants without asking crazy questions. If you know your application well, then you already know that the customer wants his or her product on time. Let's call that a given (not negotiable).

Thus, you need to ask questions of your customer as if he or she were your best friend. "What are your biggest challenges today?" is a much better question to get the ball rolling. Quality is, after all, a matter of the heart of your customer, and Voice of the Customer efforts are great and should be used, but looking at your customer's heart is much more likely to get you to a product that does not just meet basic needs, but also delights the customer in ways that will have them returning to you for future needs.

Juran [48] lists customer needs as follows:

- Stated needs—what the customers say they want (a car)
- Real needs—what the customer really wants (transportation)
- Perceived needs—what the customer thinks is desired (a new, not a used car)
- Cultural needs—status of the product (a BMW)
- Unintended needs—the customer uses the product in an unintended manner. (A BMW used to haul concrete blocks.)

13.2.2   How Is Quality Measured?

Suppose that you got some meaty answers to that heart question above. It is going to be up to you to measure your degree of success in meeting that hard-won quality requirement. When you are thinking about your customer or gathering requirements, keep in mind that whatever you determine is important in your customer's space is going to have to have some sort of handle attached to it so that you will know how to describe it in terms of concrete requirements.

13.2.3   When Is Quality Accomplished?

Parallel with the idea of measuring quality factors is the idea that you will have to define a threshold level for them so that you will know when you have found the holy grail of accomplishing the goal. If you keep this in mind as you work to build a user profile, you will have a much greater chance of actually achieving your goal. Your development program will be much more successful if you can give your team a picture of your customer in terms that make that customer real and concrete. You can back up that picture with a set of criteria and success metrics that let your team know that you have made the grade and delivered what will make your customer's heart sing.

13.3 Q-FILES

You know what they say: If quality were easy, everyone would be doing it. Well, okay, they do not all say that, but they should. If quality were simple, there would be no excuse for poor quality. You would have a copy of the Quality Bible and all the quality features and measures for success would be there. All you would have to do is dial them all up and deliver.

It is not so easy to determine what quality-isms (known here as Q-Files) are needed in each industry. For example, in the aerospace industry, there is no room for error, ever. A 1% failure rate is unacceptable.

An interesting example comes from the home entertainment industry from years ago. They conducted focus groups to determine how well a certain television (brand not important) suited the needs of the customer. Everything was discussed, but the one thing that was not extracted from those sessions was a remote-control locater. So even talking with customers and focusing on them can get you mediocre quality, but you have to be world class to delight the customer.

You can see from these examples that each group of customers is going to have a different perspective on Quality. Aerospace, health care, insurance, financial industry, adult beverages, family entertainment, communications, government, and countless other groups and subgroups are also going to have their own definitions of quality, along with things that they have not even defined for themselves but might still be useful. It is those surprising *"ah ha!"* revelations about what quality is for each group that will give your company the edge in the software world.

There also are customer needs related to the use of a product:

- *Convenience.* Technology in today's world can bring about new products and services that were not dreamed of before.

There are sections of society that limit the extent of technology in their lives (the Amish, for example).

- *Safety needs.* Products or services become available that customers need. For example, the thinning of the ozone layer is creating a greater need for sun-protection lotions and creams.

- *Product-simplification features.* New products can be complicated to use. Products or services should help ease the conversion to their use. Simplified income tax forms should help end tax filing woes.

- *Communications.* The need to be informed and to be given access to rightful information. Open-door meetings are a must.

- *Customer service.* Customers are expecting companies to have properly trained personnel on hand to handle complaints. The agent of the company should be empowered to satisfy the customer at that point.

13.3.1 More Questions

Building a good set of quality definitions and metrics will get you to some measurable results, but how do customers express Quality? Remember that this is all about getting behind your customers' eyes and understanding what they see in terms that they can relate to. "Framing" is a term in communication theory that relates to how you frame a conversation to get you and your communicating partner to be on the same page, so that both of you mean orange (the fruit) when you say "orange" (the word). That is where you have to be when you speak with or visualize your customer.

Just as you have to frame your communication model of your customer, you also need to understand how your customers express poor Quality. If you look from a marketing frame, you might say that your customers express poor quality with their pocketbooks. You are talking about quality of the product, though, so you need to frame every piece of information you get from your customer, from requirements through to simple phone calls or e-mails, in terms of what key words they might say that trigger what they think quality is not.

In their book, *Priceless* [43], LaSalle and Britton say that most managers remain stuck in a features and benefits frame of mind that focuses more on what the product or service does rather than what it offers, what problems are being solved, and how a customer experiences the product or service. When you work very hard to be able to see what the customer sees, you write some key components in your requirements documents and echo them in test plans. Often, though, you are doing contractual work, and need to be concerned about whether the measure of Quality equates to a contractual execution. Contracts are often negotiated without a single quality requirement, beyond reliability, in them. It is often the case that you will be using intuition to extract what you think are quality requirements. Again, it goes back to the idea of delighting your customer, so frame your mind from that perspective when writing or reviewing requirements contracts.

13.3.2 Mind Benders

There are many ways to access the mind of the customer. Here is a short list of tools that are common in the software industry. They are also applicable across industries. Remember, when you use any one (or all) of these tools, you have to try to access the mind of the customer.

- **House of Quality Matrix.** House of Quality or Quality Function Deployment is a tried and true way to get the voice of the customer documented. This method was used in the auto industry in the early 1980s and is in widespread use today. The downside to this methodology is that it takes time and some training to get it right. The results however, can be impressive. Spend some time reviewing what the House of Quality structure is, and determine if you can use it. You will have to weigh the trade-offs of time and energy and accessibility to the Voice of the Customer in order to determine if this method will work for you.

- **Survey Questions.** Surveys can be conducted by phone, e-mail, or in face-to-face interviews in either one-on-one or group settings. The idea is that you get access to the customers and get them to talk about their wants and needs. Once you put a face to the customer, you can then start getting a full idea of what Quality is for them by understanding the pain points and what quality is to them in their application. Of course, you could wind up with many quality attributes—more than you can deliver. In this case, you could do a card exercise in which you write the attribute down on index cards, and have your customer sort them according to their importance. Depending on how large or small your effort, you can do this manually, or you can get a statistical program that will help you get a final priority based on your customer's inputs. If you want to get the customer to be real, put down costs along with each requested quality feature to see how price might make a difference to your customer.

- **Tools to View the Customer's World.** The best tool you have to view your customer's world is your own mind. You have to look at all the inputs from your selected methods and give them the sanity test. You will still have to guard against getting caught up in group think and forgetting your original goal. Your original goal is, of course, to delight your customer by providing them the quality level that they will value. There are many templates available for free on the Internet and many consulting companies that would be happy to help your organization out. No matter what, though, you must keep your eye on your goal of staying behind the eyes of the customer and seeing what they see.

13.3.3 Constructing Q-Files

Let us assume that you have chosen your method of attack. You know how you are going to approach your customer by working on the House of Quality, interviews, or faceless surveys. You will find that there are timeless quality issues that your customer is most concerned with and then there are time-bound quality issues that may shift with time. Consider the timeless quality issues as those that you can consolidate into a set Q-file. You will want to figure out how to document these issues in a way that will allow you to call up your Q-File at any time.

One way to format your Q-Files is in the same way you document your other feature requirements. Remember, though, that your Q-Files really address timeless quality issues. A good format for this would be to have some set sheets that address each quality issue in a specific, repeatable way. You can do this via spreadsheets, depending on complexity. Our suggestion is that you keep these Q-Files as simple as is possible. Timeless quality is generally simple. Your headings may look like this:

- **Title.** Give a meaningful title, such as "Access Time," for the portal page.
- **Who.** The customer profile, such as "This is the main data entry person who is generally unfamiliar with the technical aspects of the product."
- **What.** This attribute addresses the first page on the web portal for data entry. It must be able to be accessed in 2 seconds or less.
- **How Used.** This page is used for entering into the other pages, and there are 12 functions that must have direct links from this page.

- **Measure.** Remember that any attribute that you identify will need to be understood well enough that you will know that you have achieved the goal. How will you be able to tell *from the user perspective* if you have achieved this goal? You can measure access time all you want, but if your customer perceives this attribute as being unsatisfied, then it will be unsatisfied whether or not your meet the requirement.

Delighting the customer comes only from understanding the customer's frame of reference. This frame of reference is rarely measured in seconds, but more likely to be in fuzzy terms like fast. Your job is to identify ways to make your quality attribute be perceived as being fast. If you fall into the trap of quantifying that attribute in terms that you (as a quality representative) understand, rather than in terms that the customer perceives as having value, you have failed in your quest.

13.3.4   Communicating Q-Files

The next big step after documenting your Q-Files in simple, understandable terms is to figure out how to make those Q-Files accessible to those people who need to assimilate them. There are several ways to do this. Publish them, of course, in a manual (because everyone reads manuals, especially developers!). You must think of creative ways to communicate your Q-Files. You are not only trying to reach the Quality people who have ultimate responsibility for Quality in your organization. Most organizations today do not have separate Quality organizations anyway. You want these attributes to be designed-in from the beginning. You want your product developers to understand the Q-Files intimately.

There are a couple of ways beyond publishing your Q-Files in a Quality manual to get the word out. First, you can have an open forum at which there is role playing, or have a company meeting at which some Quality representatives act as the customer. You can even make it fun by playing Customer Jeopardy, and giving prizes to the people who understand the customer. You might even decide to create a new currency at work—Q-bucks—that can be bartered for a half day of vacation or something equally appealing.

You can also build some on-line trainers by which essential information about your customer is delivered. Rather than listing all those "who, what, when, how, and whys" from your Q-Files, though, focus on what the customer looks like and how they use the product. Gather feedback from your own development community to enhance the Q-Files even further. While the development community learns about the customer community, they will undoubtedly figure out ways to enhance your Q-Files (and maybe even come up with a great new Customer Jeopardy question).

Other ideas are to build a portal on your internal website that can call up an actual face of your customer, cover your walls with your customer's face or final products, and get rid of the empowerment posters. The idea is to get that face of the customer on the ground and in front of your workers. You want them to keep this face in their minds as they build the product and make them not just understand who the customer is, but get them to feel what the customer *feels* about your product. Then when teams collaborate for process improvements, they will really understand the customer.

13.3.5 Updating Q-Files

Of course, you will want to keep your Q-Files documented and updated. Be on guard to never let them get too complicated, however.

You will probably want a gatekeeper to be in charge of updating your files. Although it will be worthwhile to get your worker's inputs on what quality features might be of interest to your customer, you do not want the landscape to get too wild and unwieldy. All efforts to build your Q-Files should pass the sanity test.

13.3.6 Measuring Q-Files

In this ideal world, where you have solicited, built, communicated, and maintained your Q-Files, how will you measure performance against those quality perspectives? What will be the incentive to keep it up? Management will want to track progress and to see if the Q-Files have enough data integrity to be meaningful to the organization at large.

Nothing seems to drive a point home more than identifying the importance of something in a performance plan. Just as you put together very specific measures of success for each attribute in a Q-file, you will want to figure out an aggregate way to measure how each person in your organization is doing against those attributes. There is no substitute for accountability, and performance plans are a great place to start.

Upper-level management will have to back the idea of measuring performance against Q-file results in order for this effort to be a true success. Additionally, if you can calculate a return on investment from your Q-file efforts, you can target an organizational number for overall success metrics.

13.3.7 Lateral Perspective

Q-Files offer a way to view your customer's world using traditional methods to gather information with a lateral perspective. You will want to be able to call up Q-Files for long-term tracking and management. Q-Files can be a way to be recognized as a company that really cares about its customer.

Build your files as outlined above, and once you have done that and you think you are finished, do some crazy stuff—go sit in with a customer, call a customer, or just go to the store and watch the customer review your product. Pull in a man off the street who knows nothing about your product and ask him questions about it. Try to see if quality personnel can define what the product is and why the quality is so great.

Never rest. Build and rebuild your files, and if you are in an industry that can share best practices, go out and network with your coworkers and share your Q-Files. If you are not in an industry that can be that open, meet with others to share ideas on getting behind the eyes of the customer. Ask how you might truly get to the heart of the customer and keep trying even after you think you have gotten it right. Your customers will love you for it and that will raise you way above your competition in their eyes.

13.4 FEELING THE CHANGE

If you have followed the examples in this book, have used the templates, and even seen some improvements, how do you get those improvements and the whole idea of CPI to stick? Sticky change is a good thing! You want your changes to be effective, but you also want them to have longevity.

In Kotter and Deloitte Consulting's book, "The Heart of Change [44], they stress the importance of see–feel–change as being key in getting a change of behavior to be realized. This reinforces what we have described so far. The "see" part is the ability to come together to identify and propose solutions to everyday problems. The "feel" part is the ability to feel the success of making the change, of relieving some pain point in the organization, and, most importantly, to feel what the customer must feel. The "change" part is what you accomplish as a group or organization. In order to effectively realize this sort of "see–feel–change" relationship, you definitely need to find some quick wins. They can even be small wins, as long as they are genuine and solve some problem. The point is that you will need to be able to bring it home to the worker in some way, whether it is in a less stressful workplace, a happier customer, or a bigger net profit that results in a bonus in the paycheck.

Figure 13.1 is an example of a possible Q-file. It is a visual way to display the information, and will help build the impression of the overall customer.

CHAPTER 13 SUMMARY

- Teamwork
  - Provide opportunities for teams to self-select
  - Trust and ease built through team exercises
  - Understand different personality types and how to emphasize strengths
  - Improve team performance with coaching and check points
- Quality with heart
  - Quality is in the eye of the customer

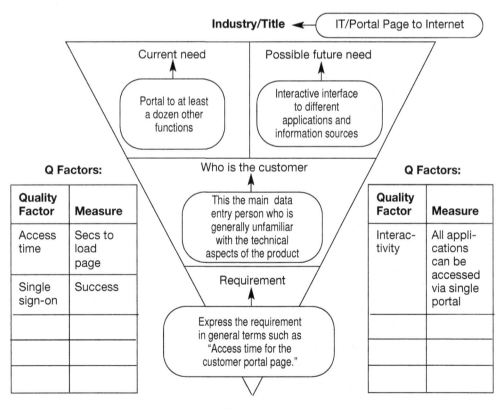

Figure 13.1.

- ○ Make quality REAL
- ○ Build the face of the customer
- Ask these questions:
  - ○ How is Quality measured? Be sure to be able to measure quality from the customer perspective
  - ○ When is Quality accomplished?
    - □ What are your user's profiles?
    - □ What is the picture of your customer? Does the customer have a face? A personality?
- Perspectives of quality
  - ○ Relating customer need to use of product
  - ○ Different industries will have different perspectives
  - ○ How is poor quality expressed?
- Mind benders
  - ○ House of Quality/Quality Function Deployment
  - ○ Surveys
  - ○ "The commonsense test"
- Constructing Q-Files
  - ○ Timeless quality attributes
  - ○ Simple quality profiles
- Communicating Q-Files
  - ○ Be creative—use open forums, role playing, Customer Jeopardy
  - ○ On-line training
  - ○ Internal portal (intranet) to face of customer
- Updating Q-Files
  - ○ Keep updated
  - ○ Maintain simplicity and understandability
- Measuring Q-Files
  - ○ Communicate to management
  - ○ Calculate return on investment where possible
- Lateral perspective
  - ○ Use Q-files for long-term tracking
  - ○ Gain varying perspectives on files; be a "secret customer"
- Feeling the change
  - ○ Identify quick wins
  - ○ Bring it home to the worker
  - ○ Communicate improvements

CHAPTER 13 AIDS—HOUSE OF QUALITY EXAMPLE

 The House of Quality, or Quality Function Deployment, is a great way for a company to understand and address market demand. There are many methods of developing a House of Quality. Some are very simplistic, whereas others are very detailed. We strongly recommend investigating which House of Quality is best suited to your needs. We used a very simple model in the following discussion.

When doing a presentation on House of Quality applications at the International Conference on Software Process Improvement in Orlando, Florida in 2006, we conducted a class exercise using automobiles. Since this was an international conference, we felt that automobiles would translate across all cultures. There were 35 students in the class, and they all immediately identified with the auto industry.

What follows is how we conducted the exercise and some of the results. We divided the class into three groups: two groups (Group 1 and Group 2) of 15 and one group of five (Company Facilitator, Customer Facilitator, House of Quality Expert, and a two-person Company Management Team). The following is the discussion we had with the individuals of Group 1 and Group 2.

> Group 1—Customers: If you were in the market for a new car today, what would be your requirements?

> Group 2—Company: Over the past five years, your company's market share has dwindled to almost 10%. Your company has launched two new models to the market in the last 12 months. Each was a complete failure. Capital funds are running very low. Another failed attempt will destroy your company. You now have one last chance to turn your company around with a new design.

Separate the two groups so that they are not able to hear the other group brainstorming. From the third group of individuals, assign one as the Company Facilitator and one as the Customer Facilitator. Their jobs will be to conduct a brainstorming session that results in five customer requirements and five company-proposed designs features. Once completed, the Customer Facilitator takes his five requirements and inputs them down the left side of the House of Quality template. The Company Facilitator inputs his five proposed design features across the top of the same House of Quality template. Now it is up to the House of Quality Expert to complete the template.

Once the House of Quality template is completed, the House of Quality Expert presents it to the company management team.

Have the company management team discuss what they have learned from the House of Quality. What follows are just some of the questions you can ask to stimulate discussion:

- Do any of the proposed design features map to market demands?
- Which proposed design features have no impact on the customer?
- Which proposed design feature has the biggest customer impact?
- Can the design feature without a large customer impact be implemented without implementing costly designs with low customer impact?

Figure 13.2 shows is the House of Quality that was created.

CHAPTER 13 AIDS. CASE STUDY—FIRST BUILD A PAPER AIRPLANE

While doing a presentation on customer facing at the International Conference on Practical Software Quality Techniques in Las Vegas in 2005, we constructed an exercise entitled *Build Me a Paper Airplane* that can vividly illustrate how customers view quality. It is a very easy exercise to conduct; however, be prepared for some very perplexed looks from your colleagues. The exercise goes like this:

- In a group setting, tell each person that you will be asking them to build a paper airplane. You should have at least two "customers." Tell the group that you and your partner are requesting that a paper airplane be built for them.
- Let people select their own paper (you should have different-colored paper, stickers, and so on available).
- Give them all a few minutes to build an airplane. You may get a few questions. If so, answer them honestly, but do not give out too much information.
- While the products (the airplanes) are being built, you and your partner should write down your requirements.

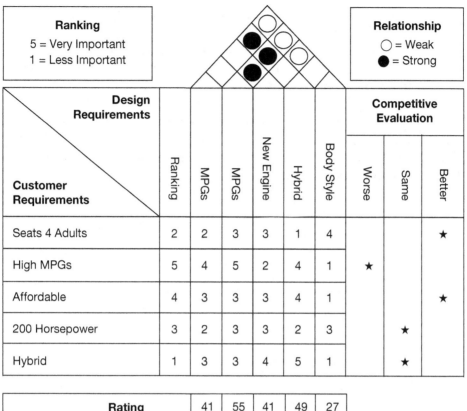

Ranking							
5 = Very Important							
1 = Less Important							

Relationship	
○ = Weak	
● = Strong	

Design Requirements / Customer Requirements	Ranking	MPGs	MPGs	New Engine	Hybrid	Body Style	Competitive Evaluation		
							Worse	Same	Better
Seats 4 Adults	2	2	3	3	1	4			★
High MPGs	5	4	5	2	4	1	★		
Affordable	4	3	3	3	4	1			★
200 Horsepower	3	2	3	3	2	3		★	
Hybrid	1	3	3	4	5	1		★	

Rating		41	55	41	49	27
		25–30	40–50	200 < HP	75% Elect	Soften

Target Values

Figure 13-2.

- After everyone has built their planes, announce that each customer is willing to trade a door prize for a plane that meets our specifications.
- Have everyone hold their planes up and, as the requirements are not met, have them lower their planes.
- Sample requirements
  - Customer 1's requirements could look like this: plane must

have at least one decal, must fit into a small gift box (6 inches by 5 inches), and a snub nose.

○ Customer 2's requirements, of course, are different. The plane must look innovative, it must have a V-shape, it must be colorful, and it must fly at least 10 feet.

The point of this exercise is that everyone thinks they know what a paper airplane is but, in reality, everyone has a different definition of quality, and that definition is likely to be different from your customer's definition.

CHAPTER 14

One World—Uniting Your Change Maps with the New World View

The voyage of discovery is not in seeking new landscapes but in having new eyes.

—Marcel Proust

14.1 CONCLUSIONS AND ENCOURAGEMENT

We are finally finished with our journey around the CPI world. Hopefully, it will take your organization less than 80 days to see some results. Moreover, we trust that the CPI principles that we have outlined here will last for your organization's long and healthy lifetime.

By using and adapting the chapter aids, we hope that you will have a successful launch. By pointing out pitfalls and potential roadblocks to process improvement, we hope to steer you away from murky waters and toward clear sailing.

Remember that adversity does not always lead to trouble—it sometimes leads to great success. Also remember that your organization is different today than it was yesterday, and it will be equally different tomorrow. This is why you must constantly adjust your efforts.

As an illustration, we decided to open our personal worlds up to you, the reader. The first map, shown in Figure 14.1, is of coauthor Jeff Fiebrich's 10-year-old daughter Caitlyn's world. As you might expect, it shows what she is most familiar with and, amaz-

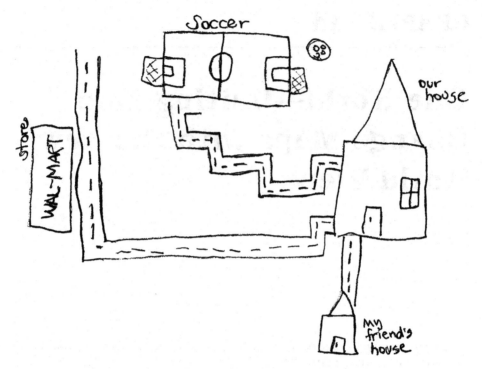

Figure 14.1. Caitlyn's map.

ingly, it accurately depicts the direction and the turns that must be taken to get around her world. You cannot tell from the map if she gets around in a car or on a bicycle or on foot, but you do understand the components of her life that are important to her and hold great weight.

Now look at the map that coauthor Celeste Yeakley's 25-year-old daughter, Libby, created. You can tell that there are very different things that are important to her and fill her life. You can tell that her child is of utmost importance to her and that her world is a bit more complicated than Caitlyn's. Libby and Caitlyn are both female, but they have totally different perspectives on life.

If you put these two maps together, they would look like they had a few things in common, such as buildings and streets and such. And you might even say that Caitlyn could use Libby's map in 15 years. The problem with that is, in 15 years the *whole* world will change, and her map is likely to be different in subtle and not so subtle ways. Either way, it takes all views to build a full world. And it takes timely maps to accurately depict today's world. The trouble is, today is always changing and only lasts 24 hours. It is for this reason that your process improvements have to be continuous.

Figure 14.2. Libby's map.

We leave you with these final thoughts. Once you have built your company's world views, put them all back together and see if you have the whole world you need to describe your customers and your own future success. By continuously pulling together everyone's thoughts, perspectives, and innovations, you will put your whole map back together into a company vision that will not only define who you are, but why you are great. If you do it right, you could rule your own corner of the world.

Definitions

This book uses the following titles when discussing personnel.

Advocate. An individual recruited by the software quality engineer. The advocate will perform the task of the software quality engineer. The advocate will be trained to perform all tasks. The advocate reports in dotted-line fashion to the software quality engineer. These tasks are in addition to the advocate's regular responsibilities.

Champion. An individual recruited by the software quality engineer and the Quality manager. Each champion is responsible for a key process area. Champions are responsible for training and resolution of issues surrounding the key process area.

Non-Quality Personnel. Any individuals that do not report in solid-line fashion to the Quality Department.

Quality Department. A department independent of the engineering group. Members of this department are responsible for the company's level of Quality.

Quality Manager. An individual from within the Quality Department, responsible for the company's level of Quality.

Quality Personnel. Individuals who report in solid-line fashion to the Quality Department.

Software Quality Engineer. The person responsible for conducting activities such as assessments, reviews, and evaluations, and facilitating process improvement meetings. Reports to the Quality Manager. Responsible for recruiting for the Advocate Program.

Acronyms

BMW	Bavarian Motor Works
CCB	Configuration Control Board
CD	Compact Disk
CEO	Chief Executive Officer
CM	Configuration Management
CPI	Collaborative Process Improvement
DM	Development Methodology
DMAIC	Define/Measure/Analyze/Improve/Control
FTR	Formal Technical Review
GE	General Electric
H/W	Hardware
HLDD	High-level Design Document
HoQ	House of Quality
IEEE	Institute of Electrical and Electronics Engineers, Inc.
ISO	International Organization for Standardization
KPA	Key Process Area
L&D	Learning and Development
LLDD	Lower-level Design Document
PDE	Product Documentation Estimate
PDP	Product Development Process
PTO	Project Tracking and Oversight
QA	Quality Assurance
QFD	Quality Function Deployment
ROI	Return on Investment
RTM	Reliability Test Matrix
SCM	Software Configuration Management
SEI	Software Engineering Institute
SPI	Software Process Improvement
SPMP	Software Project Management Plan
SPP	Software Project Planning

SQA	Software Quality Assurance
SQE	Software Quality Engineering
STMP	Software Test Management Plan
SW S/W	Software
TBD	To Be Determined
TM	Test Manager
URL	Uniform Resource Locator
WBS	Work Breakdown Structure

References and Resources

The following sources were used in the preparation of this book.

1. Fiebrich, Jeffrey D., and Yeakley Celeste L., "The Q-Files—What Is Your Customer's Definition of Quality?" in *Proceedings*, Software Develop and Expo West 2005, March 2005, Santa Clara, California.

2. Frame, J. Davidson, *Managing Projects in Organizations*, Jossey-Bass Inc., 1995.

3. Daniels, William R., and Mathers, John G., *Change-ABLE Organization—Key Management Practices for Speed and Flexibility*, Jossey-Bass Inc., 1997.

4. Gladwell, Malcolm, *Tipping Point: How Little Things Can Make a Big Difference*, Little, Brown, and Company. 2000.

5. Collins, Jim, *Good to Great: Why Some Companies Make the Leap . . . and Others Don't*, HarperCollins, 2001.

6.. Kearl, Michael, On the Needs of Selves and Societies, *www.trinity.edu/~mkearl/socpsy-3.html*, 2001.

7. Fiebrich, Jeffrey D., and Yeakley Celeste L., "Deploying Guerilla Quality—Modern Techniques for Quality Initiatives," in *Proceedings*, International Software Quality Institute 3rd World Congress for Software Quality, September 26, 2005, Munich, Germany.

8. Wright, Robert, *The Moral Animal: Why We Are the Way We Are; The New Science of Evolutionary Psychology*, Random House, 1994.

9. Fiebrich, Jeffrey D. and Yeakley Celeste L., "The Q-Files—What Is Your Customer's Definition of Quality?" in *Proceedings*, Software Develop and Expo West 2005, March 2005, Santa Clara, California.

10. Fiebrich, Jeffrey D., and Yeakley, Celeste L., "Configuration Management—A Matter of Survival!" in *Proceedings*, UML and Design World 2005—Architecture, Design, Modeling, and Beyond, June 2005, Austin, Texas.

11. Fiebrich, Jeffrey D., and Yeakley, Celeste L., "Building a House of Quality." in *Proceedings*, International Conference on Software Process Improvement, April 04, 2006, Orlando, Florida.

12. Hickman, C., Connors, R., and Smith, T., *The OZ Principle*, Prentice Hall Press, 1995, 2005.

13. ISO 9001: Quality Systems—*Model for Quality Assurance in Design, Develop-*

ment, Production, Installation and Servicing, International Organization for Standardization, 1994.

14. ANSI/ISO/ASQ Q9001-2000: *Quality Management Systems—Requirements;* ASQ Quality Press, 2000.

15. Blanchard, Kenneth, and Hersey, Paul, *Management of Organizational Behavior,* Prentice-Hall, 1982.

16. Bryson, John, Strategic Planning for Public and Nonprofit Organizations, Jossey-Bass, 2004.

17. Crosby, Philip B., Quality is Free, McGraw-Hill, 1979.

18. Westcott, Russell T., Levels of Maturity Matrix, R. T. Westcott and Associates, 2001.

19. Silverman, Lori L., *Critical SHIFT: The Future of Quality in Organizational Performance,* ASQ Quality Press, 1999.

20. Fiebrich, Jeffrey D., and Yeakley, Celeste L., "The Quality KISS—Keeping Quality Simple in a Complex World," in *Proceedings,* International Conference on Practical Software Quality Techniques East, March 2004, Washington, DC.

21. Lewis, Richard D., *When Cultures Collide,* Intercultural Press, 2005.

22. Jones, Martis, *The Prodigal Principle,* Worth Publishers, 1995.

23. Fiebrich, Jeffrey D., and Yeakley, Celeste L., "Customer Facing—It's Time for an Extreme Makeover!" in *Proceedings,* International Conference on Practical Software Quality Techniques West, May 2005, Las Vegas, Nevada.

24. Schmaltz, David A., *The Blind Men and the Elephant: Mastering Project Work,* Berrett Koehler Publishers, Inc., 2003.

25. Senge, Peter, *The Fifth Discipline: The Art and Practice of the Learning Organization,* Bantam Dell, 1990.

26. Fiebrich, Jeffrey D., and Yeakley, Celeste L., "Development Lifecycles—Plunging Over the Waterfall!" in *Proceedings,* International Conference on Practical Software Quality Techniques West, May 2005, Las Vegas, Nevada.

27. Godin, Seth, *Unleashing the Ideavirus,* Hyperion, 2001.

28. Fiebrich, Jeffrey D., and Yeakley, Celeste L., "Guerilla Quality—Innovative Ways to Engage Personnel in Process Improvement," in *Proceedings,* International Conference on Software Process Improvement, June 2004, Washington, DC.

29. Hayes, Robert M., and Wheelwright, Steven C., *Restoring Our Competitive Edge,* Wiley, 1990.

30. George, Steven, "How To Speak the Language of Senior Management," *Quality Progress,* May 2003.

31. Kaplan, Robert S., and Norton, David P., *The Balanced Scorecard,* Harvard Business School Press, 1996.

32. Chang, Richard Y., and Morgan, Mark W., *Performance Scorecards: Measuring the Right Things in the Real World,* Jossey-Bass, 2000.

33. Czarnecki, Mark T., *Managing by Measuring: How to Improve Your Organization's Performance Through Effective Benchmarking,* Amacom, 1999.

34. Ragland, Brice, "Measure, Metric, or Indicator: What's the Difference?," CrossTalk—*The Journal of Defense Software Engineering,* 1995.

35. IEEE Standard Glossary of Software Engineering Terminology, IEEE Std 729 1983.

36. IEEE Standard Glossary of Software Engineering Terminology, IEEE Std 610.12 1990.

37. Dion, Ray, "Process Improvement and the Corporate Balance Sheet," *IEEE Software,* July 1993.

38. Humphrey, Watts S., Synder, Terry R., and Willis, Ronald R., "Software Process Improvement at Hughes Aircraft," *IEEE Software,* July 1991.

39. Russell, J. P. Ed. *The Quality Audit Handbook,* 2nd ed., ASQ Quality Press, 2000.

40. Hansen, Kim, *Inquiring Appreciatively in Albuquerque,* http://www.transformativedesigns.com/appreciative_inquiry .html, March 2006.

41. Argyris, Chris, *Flawed Advice and the Management Trap,* Oxford University Press, 1999.

42. Wheeler, Donald J., *Understanding Variation,* 2nd ed., SPC Press, 2000.

43 LaSalle, Diana, and Britton, Terry A., *Priceless,* Harvard Business School Press, 2003

44 Kotter, John P., and Deloitte Consulting, LLC, *The Heart of Change,* Harvard Business School Press, 2002.

45. Pressman, Roger, *Software Engineering, A Practioner's Guide,* 4th Edition, McGraw-Hill, 1969.

Index

Printed in the United States
By Bookmasters